Development, Globalization, Global Values, and Security

Glen Segell
Editor

Development, Globalization, Global Values, and Security

Essays in Honor of Arno Tausch

Springer

Editor
Glen Segell
University of the Free State
Bloemfontein, South Africa

ISBN 978-3-031-24512-1 ISBN 978-3-031-24513-8 (eBook)
https://doi.org/10.1007/978-3-031-24513-8

This Springer imprint is published by the registered company Springer Nature Switzerland AG
The registered company address is: Gewerbestrasse 11, 6330 Cham, Switzerland

"The thematic catholicity of this volume offers a fine reflection of the width and breadth of Arno Tausch's scholarship. The collected essays address crucially important topics of our age all of which comprised Tausch's many publications. Kudos to the editors and contributors for assembling such a fine anthology celebrating a fine scholar."
—*Andrei S. Markovits,* Karl W. Deutsch Collegiate Professor of Comparative Politics and German Studies, University of Michigan, Ann Arbor.

"To shape all levels of the world so that peace will prevail—this has been the scientific and political impulse of Arno Tausch all lifelong."
—Dieter Senghaas, Professor of Peace, Conflict and Development Research (FR Germany), Author of Perpetual Peace. A Timely Assessment (2007)

"This book with contributions from experts in the field represents well Professor Arno Tausch's major contributions to the literature on development, welfare, institutions, globalization, global values, and security using large datasets and advanced composite index methodology in large cross-country comparative studies."
—*Almas Heshmati,* Professor of Economics, Jönköping International Business School (JIBS), Sweden

"I first got in touch with Arno when I was Editor of the International Journal of Health Planning and Management, and he was contributing material on how globalization was adversely affecting public health. As I came to explore his work, I realized that his recent work on health was but a small nugget in an immense quarry. This festschrift illustrates the trajectory of his work, from world systems and dependency to his studies on religions today. Early on, Arno rejected Hayek (and I wish that some of the clowns running the UK today could understand why.) His primary concern is with the marginalized and the underdog, but in the context of a morally inspired social-liberal vision for mankind. He is unusual in being a world theorist with a practical orientation to his theory at the very least in the sense that policymaking would be better-off if it heeded the

results of his research. In an age of intellectual dead-ends and self-doubt, Arno is a giant."
—*Calum Paton,* Emeritus Professor of Public Policy, Keele University, UK

"What makes this new publication—a reflection and discussion of Arno Tausch's work—especially worthwhile to read: it proves the importance of adapting analysis to changing realities. And indeed, Tausch himself always showed the need not only to look at historical developments but also to change thinking and move on in the development of methods and methodologies."
—*Professor. Dr. Habil,* Research Fellow at the Human Rights Center., Law School at the Central South University, Changsha, PR China

"As an academic in political science, reading the essays compiled in this book gave me new and interesting insights into Tausch's approach to research during his career. It is also a valuable contribution by fellow academics in these fields which builds on the work Tausch has done with the necessary mindfulness and critical analysis that gives this book valued credibility. In his introduction, the editor confidently hoped that "this work will provide a contribution, however minimal, so that the values chosen consistently show direction to our duties towards one another, our shared humanity." I believe this book does indeed adhere to these aspirations in a way Tausch surely approves of and readers will, importantly, understand. I confidently recommend this book for anyone who appreciated Tausch's commitment to these values. For some it will be a timeously reminder of our duties to one another, and for others a necessary wake-up call."
—*Dr. Ina Gouws,* Political Studies and Governance, University of the Free State, South Africa

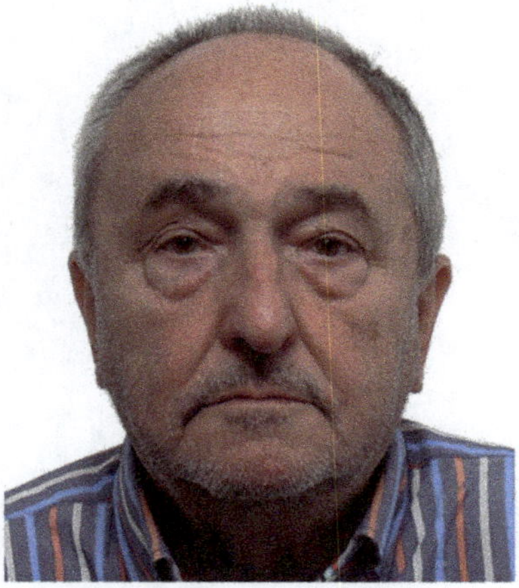

Arno Tausch

Contents

Contributors

José Alemán Department of Political Science, Fordham University, The Bronx, NY, USA

Fulvio Attinà University of Catania, Catania, Italy

Russell A. Berman Stanford University, Stanford, CA, USA

Tessaleno Devezas Atlântica – Instituto Universitário C-MAST (Center for Aerospace Science and Technologies), Lisbon, Portugal

Raimund Dietz Free University of Berlin, Berlin, Germany

Hartmut Elsenhans University of Leipzig, Leipzig, Germany

Manfred Fuchs University of Graz, Graz, Austria

Farid Hafez Williams College, Williamstown, MA, USA

Dong Wook Lee Department of Political Science, Adelphi University, New York, USA

Jacques Neriah Jerusalem Center for Public Affairs, Jerusalem, Israel

Hans-Heinrich Nolte Historisches Seminar der Leibniz-Universität Hannover, Hannover, Germany

Stanisław Obirek American Studies Center, University of Warsaw, Warsaw, Poland

Kunibert Raffer University of Vienna, Vienna, Austria

Markus Reisner Austrian Armed Forces, Vienna, Austria

Glen Segell University of the Free State, Bloemfontein, South Africa

Hussein Solomon Centre for Gender and Africa Studies, University of the Free State, Bloemfontein, South Africa

Manfred B. Steger Department of Sociology, University of Hawai'i at Mānoa, Honolulu, HI, USA

Dwayne Woods Department of Political Science, Purdue University, West Lafayette, IN, USA

List of Figures

List of Tables

Introduction

Glen Segell

Abstract Arno Tausch, the genre, and the concept of a Festschrift and the 15 authors in four sections who honored him are presented in this introduction chapter. The life and career of Arno Tausch as an Austrian diplomat, as an academic, and as a prolific author and commentator spanning over 50 years are introduced. He is known and respected for his short and accurate central essays which he wrote on four broad issues and topics of development, globalization, global values, and security that provide the framework for the sections in this volume, whilst an abstract is provided on each of the 15 chapters. An Interview with Arno Tausch is included. Incorporating evidence from published articles and books, reviews, personal correspondence, and diaries, this study honors the life and career of Arno Tausch.

1 Introduction

Let me please share some thoughts not only from the viewpoint of having to approach an international publisher Springer and not only from the viewpoint of the international market for social science literature that needed appraisal to obtain the publisher contract.

Generally, I served as facilitator, rather than editor of this volume. I must admit, though, it was almost like herding cats in a mountain range, given that the request for contributors was sent in the middle of a COVID pandemic on Christmas Eve 2021; it is a wonder that we received such a positive and overwhelming response—both quantity and quality were forthcoming. I tried to keep track of and consolidate contributions, making hardly any changes in the essays, that were all double peer-

Supplementary Information The online version contains supplementary material available at https://doi.org/10.1007/978-3-031-24513-8_1.

G. Segell (✉)
University of the Free State, Bloemfontein, South Africa
e-mail: glen@segell.com

G. Segell (ed.), *Development, Globalization, Global Values, and Security*,
https://doi.org/10.1007/978-3-031-24513-8_1

reviewed anonymously, and all filled with the legacy of Arno Tausch's wisdom and knowledge.

At the fore of the positive responses was the publisher Springer Nature Switzerland AG. The team there jumped onboard the project and have played a substantial role in ensuring the technical quality—thank you! I for one saw the opportunity to work on a Festschrift in his honor as the ideal opportunity to add to the viewpoint of the publisher (Springer) who has published many of Arno Tausch's works and to the international readership who has bought and read them.

This Festschrift is a celebration of the life work and influence of a man who is still very much alive and still thriving at 72 years of age. This festschrift is a book by different authors, colleagues, and/or students of a respected scholar. What is a festschrift? I do not ask this question for want of an Oxford English Dictionary (real or virtual), but rather out of genuine intellectual curiosity and professional commitment to ensuring that academic work continues to make distinctive and valuable contributions. Literally, the word is from the German fest meaning, "feast" or "celebration" and *schrift* meaning "writing." It is undoubtedly an opportunity to demonstrate admiration for a colleague, and this is abundantly evident in this book. As a minimum, a festschrift is a liber *amicorum*—a book of friends.

That leads to a second question—what else is a festschrift? As I write, I realize that perhaps I am asking the wrong question. Let me rephrase: what has a festschrift, become? If it is seemingly a random collection of contributions from academic friends, then this is a liber *amicorum*, but it does not differentiate the intellectual contribution of the festschrift from the ubiquitous edited book. In the same way that the edited collection must focus in and around a theme(s), so too the good festschrift must focus in and around the academic at its heart. We should be able to discern a strong sense of the person being honored's intellectual legacy. We do so with this collection honoring Arno Tausch.

To have a festschrift prepared in one's name is indeed an honor; but the converse is also true: it is an honor for certain colleagues to include their work in your festschrift. I for one am impressed by the 15 essays in this volume and stand in awe of having worked with every author who are global experts. I would like to deeply thank the contributors to this volume for the honor rests with you—every one of you. A thank you and gratitude are in order also to those anonymous peer reviewers of each chapter, of the whole manuscript and to those who contributed to the cover blurb to this work.

Focusing in to give respect to Arno Tausch who is the purpose of this festschrift I can clearly state that some of his very central essays impressed me as a scholar and as a general in reserves and that decision makers should read them for a good assessment of what is at stake in the future based upon the past.

Arno Tausch was one of the few scholars who predicted the hardening of the regime type in Russia back in 1991 or to be exact the essay which he published in Spain in 1989. These have enormous potential interest in the usefulness of Arno Tausch's work for the assessment by decision makers today. A wider range are his downloadable publicly accessible essays at the SSRN in New York (Elsevier, 2022).

Arno Tausch worked for many years in the public administration of a government ministry. You could even say that when he did, he was often approached by those elected to know precisely what the predictions he as an Austrian political scientist might have for any current world crisis. Many had not studied political science, nor did they have experience in their new role.

As a scientist working in the period from 1992 to 2016 for the Austrian Government, for sure, Tausch produced a very great number of internal reports in the field of international social policy, which also appeared in a condensed and de-classified form in scholarly journals, books, and think tank series such as the IZA Institute of Labour Economics (Bonn), the Vienna Institute for International Economic Studies, WIIW, and the Jean Monnet Chairs of the European Union https://eacea.ec.europa.eu/erasmus-plus/actions/jean-monnet/jean-monnet-chairs_en (Jean Monnet Chairs at the University of Catania, Sicily; Italy, Trier, FRG; and Wroclaw, Poland)

Tausch also published two times for the scholarly social policy journals of Austria's EU partner governments, that is the *Revista del Ministerio de Trabajo e Inmigración*, and, as it was later called, the *Revista del Ministerio de Empleo y Seguridad Social*, of the Spanish government. Arno Tausch worked with others in other countries, for example in 2009 with Alfonso Galindo Lucas of Spain (Lucas 2009) and again in 2012 (Lucas 2012). On two occasions, he published on tasks he had undertaken relating to international social policy for the Government of Luxembourg. One was a presentation for the memorial conference for Economics Nobel Laureate Franco Modigliani (1918–2003), the other was presented to the Luxembourg European Union Presidency Seminar: The Political Economy of the Lisbon Agenda, 12 and 13 April 2005, Chamber of Commerce, Luxembourg.

It therefore seemed natural for Arno Tausch to go one step further than just answering the questions as an employee but also to put his thoughts to paper for academia and the man in the street. He soon became known and respected for his short and accurate central essays which he wrote on four broad issues and topics of development, globalization, global values, and security.

These four issues and topics provide the framework for the Parts in this volume. However, titles, like labels, often do not do justice to the complexities of a scholar's work. Despite this limitation, it appears of interest to discern how policy concerns may have evolved over the years given the social contexts of the time and this is perhaps the thread or line that permeates through Arno Tausch's scholarly academic writings and work in government. The title of the sections accordingly progresses chronologically on this line. This chronology is also reflected in his academic writings—these are Tausch (1986), Tausch (2002a, 2002b), Tausch (2003), Tausch (2005), Tausch and Heshmati (2006), Tausch (2007), Tausch (2009a, 2009b), Tausch (2012a, 2012b), Tausch (2013), Tausch (2014), Tausch (2015), Tausch (2016), Tausch and Berman (2017), Tausch (2019a, 2019b), Tausch and Solomon (2019), Tausch (2020), and Tausch (2021).

Arno Tausch also was employed in academia and throughout his academic career, he also published 26 essays in conjunction with high-ranking military analysts or for geo-strategic think tanks in the following countries: Austria, Israel, Poland, Russia, and Turkey.

This Festschrift's dedication is to Arno Tausch's commitment to a socially just world where every person, everywhere can live with dignity and have his or her human needs and rights fulfilled. Only chosen values endure. Hopefully, this work will provide a contribution, however minimal, so that the values chosen consistently show direction to our duties toward one another, our shared humanity. I have no doubt that many readers of this volume have chosen to read it not only for its theoretical frameworks and the case research but will also seek to see how such assessments might be useful for the current deep global crisis which still might deepen further and further.

2 An Interview with Arno Tausch

The rest may well best be said by Arno Tausch himself and so I asked him to tell me, and he did so in a manner that I could give the title: "A look back in gratitude at the formative years of my scientific activity." This is what Arno Tausch said (and there is a longer version on the Digital Springer Web):

In this brief review of 50 years of publication activity in the social sciences, I would like to begin by characterizing for my readers the situation that presented itself to me at the beginning of my studies at the University of Salzburg in Austria in 1969. Several more appointed, other authors have clearly described the development of political science in Austria in the 1960s and the 1970s, for example Pelinka, Hummer, and König.

As the Austrian historian Pinwinkler rightly points out in his account of the history of the University of Salzburg, in the founding phase of this university it was mainly Catholic professors and professors burdened by the Nazi past who were active. At the beginning of my studies, this had changed for the better with the appointment of empirically-analytically oriented professors such as Günther Kieslich (sociology of mass communication, established in Austria as a university subject under the title journalism) and Klaus Faupel (international politics).

After completing my secondary education, in which I received a solid foundation for later university studies, especially in English, mathematics, philosophy, and post-Vatican II Catholic religion, I initially decided to study journalism, that is mass communication sociology and mathematics, but changed relatively early to the subject combination of political science and mass communication sociology and Ibero-Romance studies and economics.

Among the most important impulses I was allowed to receive at the University of Salzburg in the early 1970s were the lectures by Professor Günther Kieslich, e.g. on the methodology of empirical social sciences, and—particularly relevant—on the National Socialist propaganda apparatus, as well as the lectures from economics by Alfred Kyrer and later by Friedrich August von Hayek, who was teaching at the University of Salzburg at the time and whose entire lecture cycle I was able to attend.

Nobel laureate Hayek presented a lecture on economic policy at that time, which basically referred to his books published in 1970–1973 and was called economic policy but was basically intended to prove that there should be no economic policy by the state at all, because the markets, left to themselves, regulate everything optimally anyway. Hayek also focused on economic theory and the history of economic doctrine formation in his lectures.

When Hayek was awarded the Nobel Prize in 1974, it did not change my opinion that Hayek's approach could not convince me in any way, and during my political science courses I began to concern myself more and more with critical development and peace research in the traditions of Dieter Senghaas. It was also helpful that from the autumn semester of 1969, I began to study the regional studies of the countries of Latin America in the subject Ibero-Romance Studies. The Brazilian Jesuit priest Geraldo Cursino de Freitas, who lived from 1925 to 2001 and was rector of the Catholic University of Pernambuco in Brazil from 1966 to 1969, was one of the closest collaborators of the "red" archbishop of Olinda and Recife, Dom Helder Camara, who was known throughout the world. Dom Helder was rightly considered the central spokesperson for the progressive wing of the Latin American clergy and played an important role in the Latin American Bishops' Conference of Medellin, 1968.

It was not from Hayek, but from de Freitas that I first heard such names as Gustavo Gutiérrez, the creator of liberation theology, from Celso Furtado, and Josue de Castro, whose geopolitics of hunger stood in glaring contrast to Hayek's words on the problem of underdevelopment, which I found completely unacceptable.

In de Freitas' lectures on the regional studies of Brazil, there was already talk at that time of the destruction of the Amazon region, the marginalization and oppression of the indigenous population and the marginalization processes against the African-Brazilian population in Brazilian society. De Freitas had to leave Brazil in 1969 and never returned to Brazil until his death in Innsbruck in 2001.

During this time, I also worked part-time as the regional secretary of the Salzburg branch of the Austrian Latin American Institute and as a news gatherer for the editorial office of the current affairs service of the Austrian Broadcasting Corporation in Salzburg and at times also for the Austria Press Agency.

In 1970, a seminar on basic questions of international development was held in the Salzburg Kongresshaus, organized by the Vienna Institute for Development Studies. The Institute, with which I worked closely from 1971 to 1977, was founded by Bruno Kreisky. The Vienna Institute for Development Studies was an important think tank and served the foreign policy of the new Austrian Chancellor Bruno Kreisky. Like many Christians, I too—as Kreisky often said at the time—went part of the way with this reform movement that was so important for Austria.

The seminar, which took place in May 1970, not only brought Dom Helder Camara to Salzburg as a guest speaker, but also the Argentinean political scientist Juan Carlos Rubinstein from Buenos Aires, a prominent representative of the *dependencia* theories, that is theories that aim to explain the development processes of the countries of the global South through their dependent integration into the capitalist world economy.

Because *dependencia*, world system, globalization, marginalization, transnational integration, and national disintegration became so determinant for my later academic work, I summarize here a dispatch from the Austria Presse Agentur (APA) of 17 May 1970 the lecture by Rubinstein, which I conceived as a news collector for the Austria Presse Agentur, and which then rattled over the APA's telexes 1:1.

Rubinstein said that based on an analysis of the success of economic development in the richer countries of the North Atlantic zone and the theory that the industrial boom of these countries was only possible through the exploitation of the countries of Asia, Africa, and Latin America, one must realize that due to social injustice in the countries of the South, the actual income of the masses of the population was very low in absolute terms. The measures of the General Agreement on Tariffs and Trade (GATT) and other international organizations to reduce tariffs had not brought any substantial relief to the Latin Americans; indeed, only in the metropolises of Buenos Aires, Rio de Janeiro, and Caracas had an industrial proletariat been formed which was, however, under the total control of the backward trade unions.

Most of the Latin American population, however, was not involved in political life. Rubinstein also criticized the Soviet development; the domination of the Comecon states by the USSR was practically the same process taking place in Latin America and the intervention of the USSR in the Czechoslovak Socialist Republic (Czech and Slovak: ČSSR) was the equivalent of the US landing in Santo Domingo in 1965. Political power in East and West was exercised by power elites that were not identical with the people.

At that time, the texts on critical peace research and imperialism and structural violence edited by Dieter Senghaas were circulating among the students of international relations, and—like a sponge—I absorbed the literature cited footnote by footnote and obtained all the interesting studies by authors such as Fernando Henrique Cardoso, Osvaldo Sunkel, Pablo Gonzalez Casanova, etc., via the interlibrary loan system of the Salzburg University Library.

All these approaches seemed far more realistic to me against my social Catholic background, which already clearly emerged in my secondary school years, than the economics of Friedrich August von Hayek, who today must rightly be regarded as the pioneer of neoliberal economic policy on a global level. At that time, there were only about 10–15 students in his courses in Salzburg and Hayek was, as Pinwinkler rightly points out, very disappointed about his time in Salzburg.

In the early 1970s, those responsible for the Department of Political Science at the University of Salzburg made the wise decision to expand the range of courses in the field of political economy; to this end, they invited, among others, Joseph Alois Schumpeter's assistant at Harvard at the time, Professor Eduard März, as well as Jens Harms, who was an assistant to Kurt Rothschild, the doyen of critical political economy in Europe, at the Department of Economics at the University of Linz.

Looking back, I have to say that the lectures by März and Harms facilitated my later focus on those topics, which determined my academic life. The relations between rich and poor countries, development and underdevelopment, the long waves of the business cycle, economic inequality, the models of Kaldor and Kalecki, I learned all this for the first time in the lectures of these two teachers. What a change it was not to hear about self-healing forces of the market, but to study the works of Paul Sweezy and Paul Alexander Baran.

My first, longer seminar papers, which also found their way to later publication, dealt with world capitalist economics and apartheid in South Africa, as well as with the analysis of political systems in Latin America.

It was also during this time that my academic relationship with Anton Pelinka, who must rightly be called the founding father of modern empirical political science in Austria, began to deepen. I look back on this collaboration with gratitude, 50 years after my first publication on apartheid in South Africa and emphasize that Pelinka was not only a world-class political scientist in research, but also a gifted academic teacher and brilliant debater on radio and TV.

In 1977, I applied for the advertised position of university assistant for political science at the University of Innsbruck, which I took up after completing my civilian service in 1976–1977.

In my doctoral thesis "The Limits of Growth Theory" (first referee Klaus Faupel; second referee Alfred Kyrer) I dealt with the analysis of dependency and inequality in 40 countries of the world and with case studies for Brazil and Yugoslavia in the context of the general debate on *dependencia* theories.

Klaus Faupel, who lived from 1940 to 2016 and who was a full professor of political science at the University of Salzburg from 1973, was rightly regarded as the representative of the empirical analytical approach within the analysis of international relations in German-language political science.

As part of my dissertation project, I made extensive use of the excellent economic library at the University of Linz, and I also had the honor of participating in a seminar by the two university professors Rothschild and Laski on distribution theories at the University of Linz. In general, the academic work of Kurt Rothschild became decisive for me in the analysis of political economy.

Rothschild and Laski looked at the critical questions of political economy with modern economic, empirical methods. I liked this synthesis of a critical approach with modern, if you will, normal scientific methods very much and I do not hide the fact that in all the protracted debates in German-speaking social science between the followers of Theodor Adorno's critical theory and Hans Albert's scientific positivism, my own sympathies for Albert's empirical analytical approach were very clear.

From 1973 onward, my work for the Catholic University Community Salzburg deepened and in 1974 the Catholic University Community Salzburg organized an international seminar on the analysis of poverty in Austria under my leadership, which involved Kurt Rothschild and Anton Pelinka as well as numerous other social scientists from Austria. Then in 1977, extended by the analysis of the Project Group on International Economics at the Max Planck Institute in Starnberg (Folker Fröbel, Jürgen Heinrichs, and Otto Kreye), an anthology on poverty in Austria was published by Leykam in Graz.

Very early on, not only in my journalistic activities but also as an adult educator, I tried to bring the contributions of critical *dependencia* theories to the discussion in Austria, such as a seminar organized by the Austrian Latin America Institute in 1974 on the Feast of the Ascension, which was also attended by the well-known German sociologist Hans Albert Steeger, among others.

Since my partnership with my beloved wife Christina, whom I met in the summer of 1977 at the University of Linz, and our marriage in 1978, I also began to concern myself more intensively with the integration of the countries of East Central Europe into the capitalist world economy.

In the present anthology, the numerous topics that I have dealt with in the course of my academic work are now presented: Development theory, Public Opinion Surveys, World Values Survey, Arab Barometer Studies, Mass Communication Research, European Union—social policy, EU enlargement, EU external relations, globalization and world-system studies, Peace Research, Eastern Europe and other socialist systems in global capitalism, Studies on Global Catholicism, World Bank Pension Reforms, Muslim minorities in Europe, Kondratieff cycles, war cycles, economic growth, methodologies to study poverty, Antisemitism and Europe's Jewish Heritage, sustainable development, Political Science and Universities in the Twenty-first Century, public health research, global migration, Middle East Studies and Arab Spring, global terrorism.

On that note and not wishing to dominate I would like to leave the floor to the authors.

3 The Contributors Have Joined the Floor that Arno Tausch Vacated

On that note of "Arno Tausch leaving the floor" the contributors to this volume gave weight to Arno Tausch's work by each providing a contribution to an angle or field of his scholarship or his employment in government. A summary of each of these are as follows noting the four sections of the volume.

Before diving into these I hasten to add that Arno Tausch was modest enough to also emphasize that he has been influenced by others and pointed me in the direction of proving this. We worked together looking at the statistical evidence (Semantic Scholar in Appendix I of this chapter) to show the quantity of intellectual influence by others on his work and reciprocally his on others. From this I extrapolated, for example, that Inglehart wielded the greatest influence on his work, and Arno Tausch was among the six authors in the world, on whom Inglehart wielded the greatest influence.

In the first section of this volume that focuses on Development: Hartmut Elsenhans in his essay "Development Theory: The Social-Liberal Approach" credits Arno Tausch as presenting a coherent, social-liberal model of overcoming underdevelopment. This is followed by Kuniber Raffer in his essay "Analyzing Inequality, Dependence, and Poverty in the World System" where he credits Arno Tausch as one of the extremely few pioneers of development studies in Austria. He analyzes Arno Tausch's work as a *Dependencia*, showing that he still sees and uses dependency theory as a valuable tool of analysis, also in Eastern Europe and the EU. Arno Tausch is noted as one of the extremely few academics still showing the importance of relations of dependence. Finally, Arno Tausch's relation to the idea of Unequal Exchange is explored, a concept he has never explored theoretically, but often referred to in his writings.

Still in the same section Raimund Dietz in his essay "Politics means shaping the shape" informs that Arno Tausch, together with his teacher Elsenhans, draws a political-economic vision that he fully supports. Both say: Modern capitalism, insofar as it functions at all, does not function by itself, but when it is shaped by a good and appropriate policy. As practical, realistic, and insofar true as this statement is, the social sciences in general find it difficult to support this statement with their methodology. He then clarifies why this is so and, in all due brevity, sketches out a conceptual framework that makes it easier for political economists who, like Arno Tausch, feel responsible for the polity to advocate their good-practice-oriented strategy perhaps even better than before.

Manfred Fuchs concludes this first section in his essay "Multinational Firms and the Pattern and Structure of Foreign Direct Investment" by stating that when he first met Arno Tausch, he was a young associate professor at the University of Innsbruck in Austria in 1985. During a political science seminar, he had distributed a reasonably long reading list with reference to the world-systems theory. Arno Tausch was an energetic, emphatic, friendly, supportive, and articulate teacher who encouraged his students to be critical and self-reflective and develop a perspective that would force us to look beyond the narrow confines of our disciplines.

The second section on Globalization commences with an essay by Manfred Steger on "Global studies meets world-system theory" who in honoring the significant academic lifework of Arno Tausch, addresses the strained relationship between the rising field of global studies and the more established framework of world-systems theory/dependency theory. Arno Tausch's academic work is an instructive example of how various aspects of these two fields of global inquiry can be brought together in mutually beneficial ways. In the Austrian scholar's spirit of openness and reconciliation, this essay briefly discusses the relevance of three shared features linking global studies and world-systems theory: the rejection of methodological nationalism; a strong inclination toward trans disciplinarity; and a firm commitment to critical thinking.

This is followed by Fulvio Attina in his essay "World-scale problems and the policy response of multilateral institutions" who delves into world-scale problems addressed by Arno Tausch and the policy response of multilateral institutions. The first section reviews the existing knowledge about multilateral policymaking to address world-scale problems. The second section discusses the application of multilateral policymaking to three current world-scale problems, namely climate warming, pandemics, and mass migration.

Thereafter is Tessaleno Devezas in his essay "On the hidden pattern in the history of wars" commences with the sentence that Arno Tausch is undoubtedly a giant of the sociopolitical sciences, who contributed a lot to a modern and lucid vision of the world system that allows us today to better understand the succession of events that brought us to the current situation, which he very coherently calls as the Vortex of the Ukrainian crisis. He says that Arno Tausch is one of the most prolific authors he ever knew and with whom he had the opportunity to cooperate, and so he is honored to know that his publications may have in some way contributed to the maturing of his convictions. His publications are filled with tables and graphs exhibiting quantities, indexes, rates, etc., being very common the use of illustrated maps in which quantitative comparisons are highlighted. This essay presents an overview of the alluring relationship between K-waves and warfare, a field to which Arno Tausch has contributed significantly. Yet, in the light of this overview, it is presented an interpretation of the historical moment we are currently experiencing.

Hans-Heinrich Nolte in his essay "Russia and Germany in the system of powers" completes the full circle in context of the war between Russia and Ukraine that started in 2021 when he notes that Arno Tausch convincingly coined the basic relation between the European World-System and Russia as "treadmill"—always laboring against it, maybe successful in the military (Suvorov crossing the Alps, Russian troops in Paris and Berlin…) but never quite reaching its economic and social standards.

The third section on Global Values is opened by Russell Berman in his essay "Globalization, Islamism and the Persistence of Religion" that adds that processes of globalization examined by Arno Tausch have included both economic and cultural dimensions, whereby the latter involve secularization and universalism as parts of the long-term program of modernization. Arno Tausch's work shows that an overemphasis on Islamist extremism is as flawed as are ideological decoupling's of Islam from Islamism.

Stanislaw Obirek in his essay "Studies on Global Catholicism" contributes on his collaboration with Arno Tausch in the book Global Catholicism. Tolerance and Open Society: An Empirical Study of the Value Systems of Roman Catholics (Obirek, 2020). He did not realize then that this would be the beginning of his reflection on the global condition of contemporary Catholicism. The data gathered and analyzed confirmed the research intuitions that have been present in numerous books by sociologists, anthropologists, philosophers, and even theologians for years. The results allow for the formulation of a research hypothesis on the almost global erosion of Catholicism.

Farid Hafez concludes this section in his essay "Tausch and political controversies in Austrian media: 'Political Islam' in the Austrian Debate" by reviewing a discussion in the daily *Die Presseon* "political Islam" in Austria, in which he and Arno Tausch were involved. Discussing the main arguments of both camps that were involved, Hafez tries to answer the question, how a reasonable discussion on "political Islam" can take place in a context, where the notion of "political Islam" has been mobilized by the governing New Austrian People's Party led by Sebastian Kurz to serve anti-Muslim racism.

The fourth and last section is on Security and is opened by Jose Aleman, Dwayne Woods, and Dong Wook Lee in their essay "States of emergency: In whose interest are they invoked?" They inform that Arno Tausch's illustrious career spanned five very prolific decades. They honor his legacy with research that investigates why governments declare a state of emergency. Intending to the heterogeneity in regimes' invocations of emergencies, they follow Arno Tausch's exemplary use of quantitative analysis to answer complicated questions. They show how Arno Tausch relied on both deductive and inductive methodologies. They use an inductive approach to uncover meaningful patterns. They then conduct a non-linear analysis of state-of-emergency (SOE) episodes from 1950 to 2010. Their findings indicate that multiparty dictatorships produce significantly more SOEs compared to democracies. However, they witness fewer SOEs in the presence of more powerful executives. The latter are also more likely to trigger an autocratization episode.

This is followed by Jacques Neriah in his essay "Rabin and the Intelligence Services on the Eve of the Oslo Accords" stressing that intelligence is at the heart of decision-making in all fields, especially when it is a matter of life and death, the fate of a diplomatic initiative, a decision to start or end a war. In a similar approach, Arno Tausch wrote that during his years as a labor attaché in Warsaw from 1992 to 1999 and in the ministry for social affairs from 1999 to 2016, he never had access to classified intelligence of any source but what he did to the best of his knowledge was to provide his superiors with good and sound social scientific information which revealed to the decision makers many important points. These assessments, according to Arno Tausch, were linked with assessments what the most important think tanks had to say on a given subject, for example the migration potential of the countries of Eastern Europe to the old member countries of the European Union at the time of European Union enlargement.

The penultimate essay is by Hussein Solomon on "Global terrorism" noting that it is indeed difficult to capture the essence of a lifetime of scholarship of a scholar as prolific as Arno Tausch. This is made more difficult given his varied interests. Rather than focus on his numerous works examining terrorism in its myriad manifestations, this essay seeks to examine Arno Tausch's stance on terrorism in three areas: what drives recruitment into terrorist organizations; the relationship between Islamism, terrorism, and antisemitism; and what is to be done to minimize the scourge of terrorism? Arno Tausch's own works are contextualized within existing scholarship on terrorism. It is argued that his greatest contribution to the study of terrorism lay in his wide-angled approach, incorporating both qualitative and quantitative aspects.

The section and the volume conclude with the essay "Terror in the Twenty-first Century, The Modus Operandi of the Islamic State—What we know and what we can learn from it!" by Markus Reisner who takes look at the goals of the new terrorists as being global and aiming at a complete transformation. He draws upon Arno Tausch's questioning that developing effective counterstrategies will be one of the great challenges of the twenty-first century and of the current and future generations. The democratic constitutional state is called upon to prepare itself for the future challenges of supra-regional, global terrorism. In doing so, however, the democratic understanding of law and its basic pillars and goals, such as human dignity, freedom, justice, and legal security, must not be undermined. These are burning questions that need to be answered.

While these are the last words of this introduction there are also two electronic supplementary files. These are Arno Tausch's autobiography that is referred to in many chapters of this volume and that is entitled "Predicting the vortex: An intellectual autobiography" and a comprehensive list of literature by and on Arno

Tausch during his over half a century of political science many of which is cited in this volume. These are available at URL.

Appendix: The Scientific Orbit of Arno Tausch

Tausch is among the scholars with the highest numbers of quotations from . . .

the works of . . .	Quotations in the works by A. Tausch (according to Semantic Scholar)	Main fields of studies, which had an influence on the works of Arno Tausch	Nationality/ Main scientific activity
Inglehart, Ronald F.	342	Sociology of global values	USA
Bornschier, Volker	278	World system theory, economic cycles, quantitative world system studies	Switzerland
Jorgenson, Andrew K.	245	World system theory, global environment, quantitative world system studies	USA
Barro, Robert J.	173	Economic growth, economic cycles, religion and economic growth	USA
Korotayev, Andrey V.	146	Economic and political cycles	Russia
London, Bruce	146	World system theory, quantitative world system studies	USA
Shandra, John M.	139	World system theory, quantitative world system studies	USA
Chase-Dunn, Christopher K.	134	World system theory, quantitative world system studies	USA
Amin, Samir	125	Dependency theory	Egypt
Grinin, Leonid E.	112	Economic and political cycles	Russia
Solomon, Hussein	103	Middle East studies	South Africa
Brand, Ulrich	96	International relations theory	Germany
Weede, Erich	83	World system theory, quantitative world system studies	Germany
Goldstein, Joshua S.	76	Economic and political cycles	USA

(continued)

the works of …	Quotations in the works by A. Tausch (according to Semantic Scholar)	Main fields of studies, which had an influence on the works of Arno Tausch	Nationality/ Main scientific activity
Tibi, Bassam	66	Middle East studies	Germany
Boswell, Terry	64	World system theory, quantitative world system studies	USA
Firebaugh, Glenn	63	Quantitative world system studies	Germany
Rothschild, Kurt W.	62	Economics, income distribution	Austria
Seers, Dudley	56	Dependency theory	United Kingdom
Kentor, Jeffrey D.	51	World system theory, quantitative world system studies	USA
Raffer, Kunibert	45	Dependency theory	Austria
Moaddel, Mansoor	45	Sociology of global values, Middle East studies	USA
Tessler, Mark	40	Middle East studies	USA
McCleary, Rachel	39	Religion and economic growth	USA
Singer, Paul Israel	32	Dependency theory	Brazil
Cardoso, Fernando H.	27	Dependency theory	Brazil
Sunkel, Osvaldo	21	Dependency theory	Chile
Elsenhans, Hartmut	19	Dependency theory, Economics, income distribution	Germany
Lebl, Leslie	18	Middle East studies	USA
Senghaas, Dieter	15	Dependency theory	Germany
Kreye, Otto	13	Dependency theory	Germany
Szlajfer, Henryk	12	Dependency theory	Poland
Krasilshchikov, Victor	11	Dependency theory	Russia
Herkenrath, Mark	7	World system theory, quantitative world system studies	Switzerland

Source: https://www.semanticscholar.org/

References

Elsevier. (2022). Arno TAUSCH Author Page SSRN. Retrieved August 3, 2022, from https://papers.ssrn.com/sol3/cf_dev/AbsByAuth.cfm?per_id=782865

Lucas, A. G. (2009). La Unión Europea, La 'Ciudad en la Colina' y el proceso de Lisboa'. Revista del Ministerio de Trabajo e Inmigración, Ministerio de Trabajo y de Inmigración, Madrid. *Economía y Sociología, 77*, 69–101.

Lucas, A. G. (2012). Globalización, crisis económica mundial y desarrollo regional. Tendencias globales e implicaciones Europeas' Revista del Ministerio de Empleo y Seguridad Social, Ministerio de Empleo y Seguridad Social, Madrid. *Economía y Sociología, 96*, 147–174.

Obirek, S. (2020). Global Catholicism, Civil Society, Democracy, and Mass Migration. In A. Tausch & S. Obirek (Eds.), *Global Catholicism, Tolerance and the Open Society* (pp. 99–132). Springer.

Tausch, A. (1986). National development: Austria and Pakistan compared. In S. Farooq Hasnat & A. Pelinka (Eds.), *Security for the weak nations. A multiple perspective. A joint project of Pakistani and Austrian scholars* (pp. 245–255). Izhar Sons.

Tausch, A. (2002a). The European Union and the world system. In R. Stemplowski (Ed.), *The European Union in the world system perspective* (pp. 45–93). The Polish Institute for International Affairs, Polish Ministry for Foreign Affairs.

Tausch, A. (2002b). Европейский Союз и будущая мировая система. *Evropa, 2*(3), 23–62.

Tausch, A. (2003). Европейская перспектива: по пути к созданию «общего средиземноморского дома» и интегрированию положительного потенциала общественного развития исламс. *Evropa, 4*(9), 87–109.

Tausch, A. (2005). Is Islam really a development blockade? 12 predictors of development, including membership in the Organization of Islamic Conference, and their influence on 14 indicators of development in 109 countries of the world with completely available data. *Insight Turkey, 7*(1), 124–135.

Tausch, A. (2007). World Bank pension reforms and development patterns in the world system and in wider Europe. In S. Allegreza (Ed.), *Reforming European pension systems* (pp. 167–222). Dutch University Press.

Tausch, A. (2009a). Multikulturalität und die Armee der Zukunft in Europa - ein erster Datenbefund basierend auf dem 'World Values Survey'. *Oesterreichische Militaerische Zeitschrift, 5*(2), 557–564.

Tausch, A. (2009b). Euro-Islam. Myth or reality? A quantitative analysis based on the 'World Values Survey'. *Evropa, 26*(2), 121–151.

Tausch, A. (2012a). Globalizacja a przyszłość rozwoju i demokracji w Europie. Wnioski z nowej analizy ilościowej'. *Sprawy Międzynarodowe, 2012*(1), 3–12.

Tausch, A. (2012b). The 'European Social Model' – lighthouse for the world or sick man on the Bosporus? In S. B. Gareis & G. Hauser (Eds.), *The European Union - A global actor?'* (pp. 401–428). Barbara Budrich Publishers.

Tausch, A. (2013). A look at recent international survey data about Arab opinion. *Middle East Review of International Affairs, 17*(3), 57–74.

Tausch, A. (2014). The new global antisemitism. Implications from the recent ADL-100 data. *Middle East Review of International Affairs, 18*(3). Retrieved August 3, 2022, from https://papers.ssrn.com/sol3/papers.cfm?abstract_id=2549654

Tausch, A. (2015). Estimates on the global threat of Islamic state terrorism in the face of the 2015 Paris and Copenhagen attacks. *Middle East Review of International Affairs, 19*(1). Retrieved August 3, 2022, from https://papers.ssrn.com/sol3/papers.cfm?abstract_id=2702356

Tausch, A. (2016). Muslim immigration continues to divide Europe: A quantitative analysis of European Social Survey Data. *Middle East Review of International Affairs, 20*(2). Retrieved August 3, 2022, from https://papers.ssrn.com/sol3/papers.cfm?abstract_id=2835328

Tausch, A. (2019a). Migration from the Muslim World to the West: Its Most Recent Trends and Effects. *Jewish Political Studies Review, 30*(1–2), 1–161. Retrieved August 3, 2022, from http://

jcpa.org/article/migration-from-the-muslim-world-to-the-west-its-most-recent-trends-and-effects/

Tausch, A. (2019b). Muslim integration or alienation in non-Muslim majority countries: The evidence from international comparative survey data. *Jewish Political Studies Review, 30*(3–4). Retrieved August 3, 2022, from https://jcpa.org/article/muslim-integration-or-alienation-in-non-muslim-majority-countries-the-evidence-from-international-comparative-survey-data/

Tausch, A. (2020). The political geography of Shoah knowledge and awareness, estimated from the analysis of global library catalogues and Wikipedia user statistics. *Jewish Political Studies Review, 31*(1–2). Retrieved August 3, 2022, from https://jcpa.org/article/the-political-geography-of-shoah-knowledge-and-awareness-estimated-from-the-analysis-of-global-library-catalogues-and-wikipedia-user-statistics/

Tausch, A. (2021). Islamist terrorism, political Islam and migration in Western Europe. In H. Solomon (Ed.), *Directions in international terrorism*. Palgrave Macmillan. https://doi.org/10.1007/978-981-1

Tausch, A. & Berman, R. (2017) Measuring support for terrorism in Muslim majority countries and implications for immigration policies in the West. *Strategic Assessment* (Institute for National Security Studies, Tel Aviv). Retrieved August 3, 2022, from http://www.inss.org.il/publication/support-terrorism-muslim-majority-countries-implications-immigration-policies-west/?offset=&posts=20&type=401

Tausch, A., & Heshmati, A. (2006). Turkey and the Lisbon process. A short research note on the position of Turkey on a new 'Lisbon Strategy Index' (LSI). *Insight Turkey, 8*(2), 7–18.

Tausch, A. & Solomon, H. (2019). The demise of the Muslim brotherhood in the Arab World. *Jewish Political Studies Review, 31*(1–2). Retrieved August 3, 2022, from https://jcpa.org/article/the-demise-of-the-muslim-brotherhood-in-the-arab-world/

Part I
Development

Development Theory: The Social-Liberal Approach

Hartmut Elsenhans

Abstract Arno Tausch presents a coherent, social-liberal model of overcoming underdevelopment. In this essay, social-liberal principles are examined which combine state intervention (in areas where markets do not deliver) with market regulation (in areas where markets do deliver). This is under the headings: Globalization and internal structures; Capitalist development requires rising mass incomes; Structural heterogeneity is inevitable in underdeveloped countries and it blocks capitalist development mechanisms in the periphery. The challenge of overcoming underdevelopment, and integrating globalization into the social-liberal concept. Everything depends on the multiplier effects of additional exports. The larger the market for simple industrial products, the likelier is the expansion of the productive apparatus beyond export-oriented production. An egalitarian agrarian reform creates a minimum income in agriculture.

1 Introduction

Tausch presents a coherent, social-liberal model of overcoming underdevelopment. Underdeveloped economies are considered structurally heterogeneous. However, structural homogeneity is required for optimal operation of the market. Structural heterogeneity is a distortion of the economy under the impact of dependency on more technically advanced economies. Structural heterogeneity is a major impediment to overcoming underdevelopment. Using rent for overcoming underdevelopment requires market regulation and government intervention.

Of crucial importance is the proper embeddedness of globalizing processes into social-liberal strategies. These should aim at making economies governable by, primarily, free markets. There are two principal options for accomplishing this goal. First, state classes should invest in projects which fit into an envisioned

H. Elsenhans (✉)
University of Leipzig, Leipzig, Germany
e-mail: helsen@rz.uni-leipzig.de

G. Segell (ed.), *Development, Globalization, Global Values, and Security*,
https://doi.org/10.1007/978-3-031-24513-8_2

economic structure. In such an economic structure, rising mass incomes will trigger flexible responses within the productive apparatus. The second option comprises export-oriented industrialization. However, this also depends on rent used for currency devaluation below purchasing power parity. It must be promoted by state intervention in favor of integration industries. Overall, then, both paths to harnessing globalization in the service of economic development led to social-liberal destinations. Social-liberal principles combine state intervention (in areas where markets do not deliver) with market regulation (in areas where markets do deliver).

2 Globalization and Internal Structures

Arno Tausch presents a consistent theory of the global system and its constituent parts. One important aspect of his model is a theory of development—indeed, of overcoming underdevelopment. This is embedded in an overarching theory of globalization and general development. His model encompasses the development of today's leading capitalist countries, as well as that of the so-called transitional countries after the fall of the Berlin Wall. Arno Tausch's theory of globalization and development, including the relation between market and central planning, can be summarized as follows:

- For lagging behind economies, globalization is not necessarily deleterious. The key variable here consists in the internal structures of the economies to be globalized. Benefiting from globalization requires internal restructuring along social-liberal lines. In practice, this entails an extensive and thoroughgoing empowerment of the mass of the population as noted in Tausch (1987, 21 f., 68), Tausch and Prager (1993, 6 f.), Tausch (2003, 31), Tausch (2016a, 527), and Tausch and Jourdon (2011, 27).
- Internal economic and social structures are decisive for the impact of economic and social openness. It is vital that these internal structures are shaped according to social-liberal development principles. This is a necessary condition for the market to produce the kind of outcomes which mainstream economists expect from resources spent on accumulation (Tausch & Heshmati, 2012b, 29; Tausch, 1987, 50).
- Clearly, participants in the process of globalization require certain social-liberal internal structures in order to benefit. And yet, globalization does not spontaneously produce such internal structures. If they are absent, then globalization will only produce increasing inequality. Short-term economic growth may well be in evidence but, ultimately, such growth will only lead to blockages, and perhaps even to de-development (Tausch, 2018, 81; 2011, 511–527).

An open world economy which successfully integrates myriad participating members is superior to a closed economy in which state governments have a preponderant influence on the participating members (Tausch & Heshmati, 2013, 25ff). However, the benefits of an open world economy can only be reaped if the individual states implement internal reforms. Such reforms must, in effect, empower labor.

I fully share Tausch's view of the global system. Furthermore, I feel honored that my own contributions have encouraged Tausch in his intellectual trajectory, and in his development of social-liberal prescriptions for international political economy.

Ultimately, capitalism is characterized by the fact that capitalists are not really committed to maintaining a social-liberal system (Elsenhans, 2015, 1–27; 2022, 40–48). Capitalists strive for market imperfections, monopolies, and for limits on the expansion of mass demand, wherever this is possible (Tausch & Heshmati, 2011, 91). Exploitation and transfer of labor value is easier from underdeveloped economies. This is because the poor of the global South normally have little to no capacity to resist exploitation and unequal exchange. Only rarely do they perceive the condition for capitalist growth—that is, expanding demand.

Moreover, however, the ruling classes of the underdeveloped South are not committed to capitalist growth. Instead, they are invariably focused on maximizing financial resources. Often, they are unable to invest these resources in order to satisfy demand from competitive markets. They reject the kind of internal reforms which would free the creative productive forces of their societies. On the contrary, they prefer to ask for transfers from the developed North, thereby increasing rents in North/South relations (quoting Myrdal, Tausch, 1987, 1).

We are thus faced with a conjunction of, in the North, monopoly-seeking companies and disinterested political powers and, in the South, relatively small rent-based ruling classes. Perhaps it is unsurprising that the impulses toward growth triggered by globalizing forces tend to peter off. Here is the connection between Tausch's contributions and the work of Bornschier. Like Bornschier, Tausch observes historically limited growth trajectories which end in stagnation (Tausch, 2018, 81; Tausch & Heshmati, 2011, 91).

Tausch systemizes these ups and downs in a theory of different types of growth and crisis at the global level. He privileges Kondratieff cycles, which coincide with the exploitation of new discoveries and inventions realized at the end of each cycle (Tausch, 2016b, 68). The availability of such innovations allows for substantial spurts, such as massive simultaneous investment. Historical examples of such spurs include the Industrial Revolution in Britain, the so-called second Industrial Revolution of steel and petrochemical products, the automobile revolution, and the contemporary microelectronic revolution (Tausch, 1991, 18).

The social-liberal model of development requires high levels of employment. It depends on demand, rather than supply, as the key driver of growth. Rising effective demand raises employment. The mechanisms of growth are technology and education. Household savings do not promote growth, but government spending does (Tausch & Heshmati, 2012a, 19).

Arno Tausch thus advances a model of social-liberal—and thus socially acceptable and humane—globalization. He applies this model to the quantitative analysis of internationally available indicators. The competitive advantage of his work consists in the analysis of quantitative data.

The data are indicators of achievements in particular policy fields. This tends to privilege achievements which are (relatively) easily measurable. The data are valuable and describe achievements which are relevant for further development, such as education, demographic growth, gender equality, and income distribution (or, at least, poverty ratios). Let us assume, as I suggested above, that internal structures can enhance political influence over economic development. This being the case, then indicators of democracy, or of values favorable to democracy, can be added. This remains so even if their direct impact on economic transformation is difficult to evaluate.

Such indicators allow for an evaluation of achievements in the transformation of economies and societies. I will show here that typical blockages of development result from mechanisms which do not show up in the international data collections. The impact of impulses toward economic growth varies according to the level of integration of different branches of a given economy. The degree of integration can be calculated from the social accounting matrices which developing countries establish in order to rationalize investment in development and economic growth. I am not aware of the availability of international comparisons of the links which can be triggered by development investment. Certainly, such links do not depend solely on observable factors, but also on the flexibility and innovativeness of the so-called informal sector. For this second realm, the data are incomplete. In Asia and North Africa in the 1980s, I observed dynamic car repair shops which were capable of producing metal parts with 100th mm of precision. In sub-Saharan Africa the level of competence in metalworking was much lower.

General indicators such as education or gender equality indicate a push toward modernization. This can be considered as linked to development, and thus to the growth of productive forces. That said, such factors may result from rent-financed employment and income promotion based on the politically rational concessions of rent-based ruling classes.

Several other indicators point to the degree to which the globalization input is embedded in the economic structure. These include demography, gender equality, and share of industrial production. However, the indicators do not suggest which mechanisms might facilitate an improvement of these factors.

Embedding globalization in the social-liberal structure requires high linkages between foreign economic impulses and locally available economic structures. In the case of underdeveloped economies, these local economic structures are invariably deformed. The internal process of embedding can be defined as the provision of economic structures which favor high multipliers. This might be accomplished by market or by nonmarket mechanisms.

Structural homogeneity is characterized by low variability in productivities of factors of production in different production lines. Structurally homogeneous economies react flexibly to market signals. Structural heterogeneity is characterized by high variability in productivities of factors of production according to production lines. In such cases, price signals do not trigger flexible adjustments.

I will show here that structural heterogeneity is the result of a specific mode of integration into the global economy. I sympathies with Tausch's emphasis on markets and his mistrust of the state (Tausch & Prager, 1993, 147). Embeddedness is guaranteed if there is structural homogeneity. In turn, structural homogeneity is ultimately achieved by expanding mass incomes. This is because, with few exceptions, the impact of impulses from the global market is too weak to trigger the process of mass production through mass consumption. Precisely this process is characteristic of capitalism.

This is precisely why I fully understand and appreciate Tausch and Prager (1993, 3 f.) who reference the reformist branch of social democracy in the 1920s (in his case, the Austrian branch). These thinkers argued that capitalism needs social reform.

3 Capitalist Development Requires Rising Mass Incomes

The social-liberal modernization of capitalism must be grounded in three fundamental theoretical lynchpins. These consist in (1) a theory of profit, (2) a theory of the relation between accumulation and productivity, and (3) a theory of the realization of profit (i.e., the profitable sale of an increasing production).

All three of these theoretical lynchpins are closely connected. They carry important consequences for the functioning of the capitalist system at both the global and the national levels. In this respect, I would especially emphasize the absence of an increasing capital-output ratio in mainstream terms, and an increasing organic composition of capital in Marxist terms (Elsenhans, 2022, 71–78). If productivity increases are not linked to an increase in the capital-output ratio, then the Heckscher Ohlin Samuelson theory of international trade is clearly rooted in a myopic understanding of the logic of capitalist growth. Instead, we must turn back to Ricardo. There are significant implications here for the modalities of integrating laggard economies into the global system.

Let us begin with the origin of profit. There is a fundamental opposition here between pre-Keynesian and Keynesian economics. Ultimately, profit is not a residual which capitalists can appropriate if they are able to apply sufficient pressure to labor. In fact, profit results from capitalist investment spending—irrespective of the productivity of this investment spending (Kalecki, 1942). Let us assume that capitalists in the consumer goods industries want to sell their products at higher prices than the direct and indirect costs of production. Here, indirect costs represent the costs of inputs and worn-out equipment. In such a case, it is essential that there exists a pool of consumers whose incomes originate from beyond the consumer goods industries.

For the purposes of simplification, I assume that consumer goods are bought by households which receive wages. For their managerial activities, managers also receive wages which are additional to their access to profit. All wages—but only wages—are spent on consumer goods. In this simplified demonstration, no saving out of wages occurs. The state sector is assumed to be in equilibrium, with all government costs being paid by taxes on the private sector. Moreover, we disregard foreign trade, with exports and imports assumed to be in balance. Let us assume that there are no investments beyond worn-out capital goods. Net investment is zero. Consequently, consumer goods producers cannot sell their total product for more than the incomes they have paid as wages, directly or indirectly. In order that producers can sell their goods at higher prices than incurred during the production process, there must be a demand rooted in the wages not paid by the producers of consumer goods.

In this simplified scenario, such additional incomes can only originate from investment goods industries. Capitalists in consumer goods industries make a profit only in case of net investment spending. There must exist production with wage incomes in investment goods industries beyond the cost of replacement of worn-out investment stock (Elsenhans, 2022, 40–45).

Investment goods producers are familiar with all the technologies used in consumption goods production. Any innovation in this sphere requires new machinery. It thus necessitates a transfer of the consumer goods producer's innovative knowledge to his supplier in the investment goods industries. In this way, investment goods producers can enter consumer goods production—if, of course, the profit rate is higher than in investment goods production. A positive profit rate in consumer goods industries implies a positive profit rate in investment goods industries, and thus the necessary prices for investment goods to yield such a profit. Clearly, then, positive profits in capitalism depend on positive net investment.

At constant wages, any technology reduces unit costs (von Bortkiewicz, 1907, 456–459). The introduction of new technologies must reduce overall unit costs, irrespective of how expensive the new technologies are. They replace human labor by machinery. Technology-intensive innovations may be present but, in relation to previous technology, they save more labor than their higher cost in capital goods (Elsenhans, 2022, 48). New technology which increases machinery costs without decreasing corresponding labor costs is known as "defensive investment." This occurs when rising real wages increase the wage bill of an old technology. In such cases, rising wages render the labor cost of the old technology uncompetitive when compared with the lower labor cost of the more technology-intensive technology.

The introduction of new products which do not replace older products necessarily follows the same rules of reducing unit costs. Moreover, the extension of the production of already known products is dependent on rising demand. If capitalists are exposed to competition, they are also collectively dependent on rising real wages.

Capitalist economies thus depend on rising mass incomes. However, this is by no means to imply that capitalists intentionally create the empowerment of the masses. If they are not checked by political power, then the capitalist mechanism of appropriating surplus on anonymous markets leads to an underconsumption crisis. In the face of such a crisis, capitalists inevitably transform themselves into rent-collecting political exploiters. This is convincingly shown by the withdrawal of monopolistic long-distance trade capitalists into political oligarchies. Venice and Bruges are two prominent historical examples—and certainly, there is no shortage of such examples.

Rising mass incomes thus effectively guarantee the sale of an increasing production because any innovation reduces costs in relation to product. Increasing demand (more or less) guarantees high levels of employment, because capitalists are interested in satisfying market demand. Rising mass incomes and high levels of employment are the basis for the constitution of a working class.

Let us suppose that an innovation takes place in each production line. This will lead to an above average profit rate. Capitalists in this production line will expand production because there exists no investment opportunity which could be more profitable. If employment is high, then expansion of production means that higher wages must be offered to workers in order to lure them away from other branches. Production lines which thereby lose labor can maintain their volume of production only by raising wages, irrespective of whether they experience higher levels of production. They must raise prices. The upshot of all of this is that, in Pakistan, a haircut costs one/twentieth of the price that it costs in Germany, despite broadly comparable levels of productivity and quality.

In capitalist countries, there will typically be considerable divergence in productivity increases between different production lines. Price movements will occur. The highly innovative production line will experience saturation, and it will inevitably attract labor with higher wages. Conversely, the non-innovative production line will raise prices. As a result, the value of production which is required in order to cover the cost of labor, inputs, and capital will converge per factors of production engaged. Productivities as value of production in relation to factors engaged will converge. This remains the case irrespective of the differences in productivity increases in physical terms. Structural homogeneity is a social product. It does not depend on technical factors.

The convergence of wages in cases of high levels of employment has obvious social implications. For one, it results in large masses of salaried labor experiencing similar conditions of life. This leads to the constitution of a relatively homogeneous labor force which develops common convictions—that is, class consciousness—and thus organized movements of solidarity. Only capitalism can give rise to a widely shared consciousness of being in a homogeneous situation. One element of this consciousness is the conviction that workers under capitalism produce a surplus. This adds an awareness of being able to simply withdraw from work, to make a material impact on the system, rather than simply being able to express one's opinions.

Homogeneity of the exploited class also implies the possibility of collective action of a very gradualist type, such as temporarily withdrawing from work by striking. Workers who do not produce a surplus cannot hurt the capitalist system by striking.

This homogeneity of labor is "artificial." It is not the result of a harmonious economic or social development. It is caused by a process of convergence which is triggered by high levels of employment, despite highly heterogeneous increases in productivity.

There is no reason to assume that productivity increases are particularly high in production lines which combine a high value of capital with a relatively low amount of physical labor. Technical development is independent of the capital intensity of production. Highly innovative branches, such as office machinery or machine-tool production, have a very low capital intensity per job.

The capital-output ratio is high in mechanized industries, such as assembly line production. This is even though physical productivity increases may be low. The capital-output ratio is low in machine building or office machinery production.

Any technology must reduce unit costs of production. According to this rule, the increase in capital per unit produced is necessarily limited. An increasing amount of capital must lead to increasing levels of production. The Marxist tendency of the fall of the profit rate cannot occur. A more capital-intensive technology will only become profitable if there is a minimum increase in net output (Elsenhans, 2022, 48). This corresponds to the stability of the capital-output ratio in capitalism over the last 130 years—that is, during the period for which we have data (Helmstädter, 1969; D'Adda & Scorcu, 2003).

Capitalism is characterized by the absence of an increase of the contribution of capital to production. To be sure, the data show a rise of capital in relation to the number of workers. But this is misleading, even if it is presented in constant prices. Capital goods are produced by labor. Indeed, they are produced by increasingly expensive labor if we assume that real wages increase. The relation between the amount of labor embodied in the capital stock in relation to the production which can be achieved with this capital stock can be estimated based on a comparison of the value of the capital stock to the value of net product. This can be calculated either in constant prices or in current prices.

The relation remains remarkably stable. Capitalism does not grow because increasing amounts of indirect labor (that is, labor employed in the production of investment goods) combine with decreasing amounts of direct labor (that is, labor directly employed in the operation of these investment goods). Capitalism grows because of technical improvement. Such improvement is not necessarily capital-absorbing.

In physical terms, technological progress is achieved by benefiting from economies of scale and larger production runs. Large internal mass markets or high competitiveness on the world market are factors which allow for the promotion of technical progress. Rising mass incomes reduce the theoretically available financial surplus. This would otherwise float around freely, and ultimately find a home on financial markets. At the same time, rising mass incomes increase the possibilities of profitable investment in the real economy by expanding demand. They expand demand for those products in which technical progress is achieved by economies of scale.

Economies of scale, by the way, have an important influence on luxury products. They render feasible, or even conceivable, products which previously existed only in fairy tales: from the bicycle to the car, from the car to the aero plane, from the aero plane to the spaceship.

4 Structural Heterogeneity Is Inevitable in Underdeveloped Countries, and It Blocks Capitalist Development Mechanisms in the Periphery

If modern technology is combined with cheap labor, then there are very good grounds for assuming that pure market regulation is inappropriate. Despite this, most developing countries are currently dominated by precisely this constellation.

If productivity increases do not depend on the increase in capital per labor or the capital-output ratio, then the Heckscher Ohlin Samuelson neoclassical theory of foreign trade cannot be credible. This is partly because, in the leading industrial countries, some labor-intensive production lines may have higher productivity advances than in more capital-intensive production lines.

This is the so-called Leontief paradox (Leontief, 1954, 37). The USA has exported, and continues to export, labor-intensive products, which require high skills from the labor force. The USA also imports capital-intensively produced products. This applies not only to capital-intensively produced raw material imports, but also to some manufactured goods. Let us set aside, for a second, classic labor-intensive products such as clothing (which, incidentally, are also labor-intensive in the capitalist countries, as indicated by high fashion production). Beyond these spheres, higher productivity depends on capital-intensive technologies. Consequently, newly industrialized countries of the global South are rendered competitive in capital-intensive products (as noted as early as 1975 by Boatler, 506).

Less capital per worker cannot make labor more competitive in branches of production in which the lag in relation to leading industrial countries is high. This is despite the low capital intensity of production processes in such branches.

Laggard economies evince different degrees of backwardness in relation to leading economies. However, these differences are not related to capital intensity of production; they simply reflect differences in lags without their being grounded in factor proportions.

The structure of technical progress, especially its independence from the degree of physical capital of production technologies, wholly discredits neoclassical foreign trade theory.

A clear conclusion follows from this basic disconnect between technical improvement and capital intensity. In short, adjusting international labor costs by devaluation, whether deliberate or imposed by terms of trade deteriorations, does not lead necessarily to spontaneous increases in competitiveness in new specialisations, with relatively rapid expansion of employment and foreign trade income. There exists an obvious opposition between high foreign trade incomes and increasing levels of employment.

Complete specialization does not depend on following factor proportions; it depends on currency devaluation. This must proceed up to the level where the decreasing marginal product in potential export sectors is covered by earnings on the global market. This amount may be lower than the inevitable cost of reproduction of labor. Perfect specialization is not possible. Moreover, any attempt to approximate perfect specialization by devaluation may entail a loss of foreign exchange. Alongside internal inflexibilities, this is a major argument for rent appropriation.

Moreover, any adjustment of the productive apparatus according to the global market price system will not necessarily provide openings for increased employment. In fact, the global market price system will provide lagging production lines with high incomes. Due to the inflexibility of the rest of the economy as a result of its laggardness, these may manifest as high demand for imports. If high incomes are not spent locally, then they do not contribute to the additional demand for local consumer goods industries. High earnings from relatively productive branches do not trigger local consumer goods production. Nor do they launch local investment goods production, especially not if lags in investment goods production are high. Consequently, the high-earning export branches order their equipment from the world market.

Overall, high incomes in some branches do not become demand for local consumer or investment goods industries as a result of market incentives. Under such circumstances, the Kaleckian mechanism cannot operate. High incomes in some highly productive branches can contribute to investment, provided that this investment is realized without the perspective of being profitable even after a delay.

Because of the structure of capitalist technical progress, the Kalecki model does not work in an underdeveloped economy in which productivity lags in relation to the leading industrialized countries diverge considerably between production lines.

I will now illustrate this argument by pointing to the standard reaction of underdeveloped countries when faced with deteriorations in their terms of trade. Lewis (1978, 13) has shown that laggard economies must accept such deteriorations. They are forced into this acceptance if locally produced substitutes for the imports paid through export earnings prove to be much more expensive. Let us imagine that, for a particular economy, sugar exported to the global market buys twice as much grain as can be obtained from the soils used for this sugar production. In such a case, the local production of grain only becomes economically rational if the terms of trade are reduced by more than 50%.

The standard description of the demand price function posits a particular formula. In the standard version, any decrease in international income due to local labor is matched by a complementary increase in employment. In such a scenario, the total earnings of the respective economy do not change. In fact, because the international price of labor is decreased, local employment increases.

Let us suppose that there exists a substantial lagging behind of productivities of different segments of local labor. This can be captured in a simple demand price function: $x = -3\,y + 6$. Total sales are 2.25 at $y = 0.5$, 3 at $y = 1$ and 2.25 at $y = 1.5$.

Over the last 100 years, it has not been unusual for underdeveloped countries to experience decreases in their terms of trade in the range of 40%. Variation is high even for a country so successful as South Korea. This is seen from World Bank: World Development Indicators (2022), for example the Terms of Trade of South Korea are as follows quoting from this source where the year 2000 is the base line of 100 so in 1980: 114 (p. 3), in 1982: 117 (p. 4), in 1984: 119. (p. 6), in 1986: 131 (p. 6), in 1988: 138. (p. 11), in 1990: 133 (p. 13), and in 1995: 138. (p. 15).

Let us turn to the availability of foreign exchange. In cases of structural heterogeneity, it might not be advisable to devalue to the level where each segment of the locally available labor force is internationally competitive. Imagine that the oil-producing countries stop restricting quantities and open their oil wells unreservedly. Prices would fall to the level of costs. Rent would disappear. The redistributive mechanisms in favor of the affluent would cease to function, and no more crumbs would be thrown to the poor. Even the strongest critics of oil-producing rentier economies would quail before recommending such an outcome.

For those economies which fail in becoming structurally homogeneous, the nature of capitalist progress implies the availability of rent as a source of finance. This further points to the danger of neglecting the necessity of using rent for overcoming underdevelopment (Singer, 1950, 482). It also indicates the political danger of polarization. In such economies, the apex of the power structure is invariably occupied by a rent-based ruling class. In some cases, this ruling class is benevolent—but only rarely. Above all, this class is capable of dominating society through clientelist relations of patronage.

Akamatsu has been an important theoretician of the transfer of homogeneity within capitalist systems in which the participants are all structurally homogeneous, despite their existing lags (Tausch, 2014, 18, 185; Tausch et al., 2017, 8). In most cases, such a cycle emerges based on an innovation. First, laggard economies accept specialization. Following this, they proceed to import substitution of those products which had previously been imported. Ultimately, they develop local production of the technologies which are used for these products. This is said to lead to saturation and economic decline, until a new set of products emerges. Overall, the Akamatsu model stipulates a fundamental decline of economic activity through imitation, and the subsequent appearance of innovations when economic activity declines at the end of each cycle. If we assume a constant rate of technical innovation, then such a process should be the natural result of technical progress. There is an excess of financial resources due to saturation, which limits investment, whereas innovation continues at the previous rate. In this way, new investment opportunities appear.

5 The Challenge of Overcoming Underdevelopment

I have defined underdeveloped economies as structurally heterogeneous. This term is not meant to imply anything about the cultural characteristics of peoples or their economies. This is also noted in Córdova (1973, 63), Elsenhans (1996, 94 f) and an overview on the concept is in Nohlen and Sturm (1982, 94).

I draw on a strictly economic definition of structural heterogeneity: as differences in productivity, which are measured in prices that are practiced on the world market (Elsenhans, 1996, 95). This occurs between individual manufacturing production lines, not just between major sectors of the underdeveloped economies. In most data bases, there is no differentiation according to production lines within manufacturing and this is noted in the 10-Sector Database (Groningen Growth and Development

Centre, Timmer et al., 2015). The same lack of differentiation applies to other databases such as the CEPALstat, UN Monthly Bulletin, or EU Klems Project.

If the Western world disappeared overnight, there would obviously be no remaining differences in productivity lags with respect to the underdeveloped versus the developed world. Production costs in the more advanced underdeveloped economies would determine the prices in a newly constituted world market. If all underdeveloped economies were similarly backward in relation to the absent West, then the Global South would come to resemble the Global West of the nineteenth century. It would converge, as in the Akamatsu model.

Clearly, however, the high incomes of the highly competitive production lines constitute a financial resource. This can be neatly deployed in order to restructure the economy with nonmarket mechanisms. If rents can still be appropriated, the implied solution of totally withdrawing from the world market is not advisable.

If the aim is to draw on the growth mechanisms of the Western world, then mass production for mass consumption must be availed of. Let us assume that available openings in the world market are not sufficient for absorbing low productivity labor, especially labor which produces less than its cost of reproduction in the agricultural sector. In such a scenario, internal redistribution is necessary. In order to broaden the effect of exports, multipliers must be optimized. Local investment goods production as the basis of profit requires orientation to mass production, and thus a subsidized entry into export production lines (Elsenhans, 2004).

The more that market incentives can be used for this reorientation of production, the better for the implementation of such strategies. However, I assume that, without complementary nonmarket mechanisms, the absorptive capacity of the world market is insufficient for achieving the absorption of marginal labor. Ultimately, this is rooted in the intelligent use of rent by (hopefully) slender state apparatuses.

I share Arno Tausch's skepticism with respect to the state classes and their commitment toward development (Tausch & Prager, 1993, 7; Tausch, 1989, 146ff). Indeed, I have written extensively on this topic. I do not deny that, in some countries, state classes have made notable accomplishments, from China to India and Vietnam to Thailand. Even in some oil countries, such as Algeria, there were significant achievements. Overall, though, the relevant question is not whether we personally like the state classes. The question is how the state classes can help their economies maintain a course which is favorable to overcoming rents by using rents.

The idea of planning investment for a somewhat coherent economic structure is indelibly linked to the name of Leontief (1951). The procedure consists in devising an intended structure for the economy from which the deficits of the existing economy can be deduced. In this way, precise projects of investment can be defined based on the linkages which have to be established between the different branches of an economy. Earlier in my career, I was unfamiliar with the practice of socialist planning. I came across this model in my analysis of French colonial policy during the Algerian War of Independence.

From 1955, successive French governments faced generalized insurrection in Algeria. They grasped for any intermediary political solution short of granting Algerian independence. This included economic and social reforms, which were intended to satisfy the desire for material improvement which the Algerian masses associated with political independence (Elsenhans, 2000, 640–648). These measures were designed to generate acceptance of political solutions oriented toward creating political and legal equality. There were, however, restrictions. These guaranteed that Algerians would not be able to instrumentalize institutions they dominated. This, it was feared, might culminate in the launch of an evolutionary political process toward independence (Elsenhans, 2000, 781–796).

The late-colonial economic and social reform policies of France in Algeria constitute the major and perhaps only case in which a Western power was ready to commit important resources for overcoming underdevelopment. Indeed, France mobilized considerable financial and human resources for achieving this goal. The project was launched by young economists who were inspired by Leontief. In 1958, they presented a ten-year development project. This later constituted the first phase of the Gaullist decolonization of Algeria, otherwise known as the Plan de Constantine.

At this time, the French market was totally accessible to Algerian manufacturers, while government subsidies rendered newly created Algerian enterprises competitive on the French market. Because of these favorable conditions, little thought was given to the creation of an internal mass market through income redistribution. Redistribution ultimately hurt the European minority in Algeria—that is, the same group whose interests were supposed to be represented in the Plan de Constantine.

The plan failed because the French government opted for Algerian self-determination. This was the entirely predictable outcome of a referendum. The overall result was not the much-feared hostile secession of Algeria, but Algerian political independence. This, it was hoped, would be mitigated by continued and extensive Algerian French cooperation.

In the dying phase of the Plan de Constantine, it became increasingly clear that subsidies within the overall structure of a common market would not be sufficient. Clearly, stronger measures of dissociation, such as customs barriers or currency devaluation, were required.

From 1966, a new Algerian development policy was initiated, with the first pre-plan beginning in 1967–1969 and two subsequent four-year plans thereafter (Elsenhans, 1977, 16–31). These initiatives drew on many aspects of the first Plan de Constantin.

The plans encompassed two major innovations. First, a radical agrarian reform was meant to strengthen mass demand. It attributed a quarter of Algerian agricultural land to self-managed estates, equalizing distribution while maintaining technical efficiency. The allocation of overall investment to production and specific projects into an industrial economic table was no longer pursued. Instead, this problem was solved through Destanne de Bernis' doctrine of industrializing industries. This entailed an intensive concentration on the production and processing of newly discovered Algerian oil and gas reserves. These were to be promoted and cultivated by the state (Destanne, 1966; Destanne & Ghozali, 1969). The idea was to sow hydrocarbon rent into productive projects (Destanne & Ghozali, 1969, 262–266).

From this point, I had two major points of contact with this strategy. First, on a theoretical level, the vision of an independent Algeria was rooted in Marxist theory. It basically adhered to the model of non-capitalist development (Kramer, 1967, 1395; Marung, 2017, 29). At the core of this model lay the idea of socialist accumulation. This was to be realized in the form of government investment in industrial production for the creation of productive capacities. The statistical office collected data about household spending, which was intended to orient decentralized investment. However, the Algerian administration was unable to use this information as a guide for future economic projects.

These data suffered from a serious flaw: they did not distinguish between qualities of products according to the income strata of different consumers. Such products appear to have the same use value, but they imply very different technologies and inputs. It was not possible to test the hypothesis that poorer households consume cheaper products. Apparently, such products are technically less demanding. They allow for simple, locally accessible technology to be used, and they provide the basis for launching local equipment production. However, it was possible to test the sectoral demand of households in variegated total spending categories.

The absence of a sufficiently detailed analysis of household spending made it easy for Algerian intellectuals to maintain their focus on maximizing accumulation. It also allowed them to focus investment on production lines which were upstream and downstream of hydrocarbon production. Space was provided for an existing, and potentially expanding, private sector. This served as a potential competitor for scarce investable resources, though the private sector's possible role in satisfying mass needs was not considered.

This connection on the theoretical level to the Algerian development process was complimented by a more practical relationship. Some Algerians, in particular the Algerian ambassador to Bonn, showed some interest in my early work in this area. In effect, this entailed a radicalization of the original approach of the French Perspectives décennales. The idea was to use the internal mass market for achieving industrialization. This was to be achieved by playing on two mechanisms. Larger production batches were foreseen. This was partly due to the greater homogeneity of demand among poorer households as compared to richer households, as well as the lower technical requirements to produce products consumed by poor households in comparison to those consumed by richer households.

Poor households consume a smaller range of products, and in larger quantities. Initially, the products they consume are relatively simple. They allow for increasingly complex products and the launching of standard local investment goods industries. These permit perfection of mechanical production through increased learning by doing. There is an Algerian description of the project (Modèle d'industrialisation par la P.M.I. 1985).

I launched an enquiry into 250 small-scale enterprises on the outskirts of Algiers. I undertook this project in association with the Ministry of Planning, and with the Institute Ce.N.E.A.P. We enquired into the productive capacities and the equipment of these small-scale enterprises.

Our enquiry was to be complemented by a further investigation into the availability of modern consumer durables in Algerian households. We also wanted to know something about perceptions of technical alternatives to small-scale entrepreneurs. Households were very reluctant to reveal the availability of consumer durables. Similarly, small-scale entrepreneurs had no knowledge or awareness of technological alternatives, such as simplified technology which could be locally produced. They believed—perhaps not incorrectly—that the technology they used was already simple, although not locally produced.

However, we conducted different simulations, one of which assumed a level of income distribution comparable to Norway's. In such a scenario, consumer durables and production lines with a high level of external technical dependency (such as cars) experienced declining demand. There was an increase in household equipment, housing, some electrical consumer durables, clothing, health and hygienic articles, and communication and transport. The demand for mechanical industries increased considerably (Elsenhans et al., 2000, 208–212).

There can be doubts around flexible adjustment, especially with respect to the local investment goods industry. That said, our experience with the Algerian car repair shops and spare part producers for the textile industry showed reserve capacities in the mechanical industries. We also encountered quality standards which corresponded to the requirements for spare parts in car repair and in textile machinery.

Unhappily, the Algerian state class had little interest in launching small-scale producers with appropriate equipment. The intervention of the head of the ministry was often necessary for the continuation of our research. Nonetheless, we demonstrated that an alternative distribution of incomes would constitute an important basis for industrialization. Rising mass incomes emerged as a driver for industrial production.

We did not demonstrate that Algerian mechanical industries would follow the change of demand. For other industries, we were relatively optimistic. The technologies which we found in these small enterprises were simple, and they could be considered accessible to modernized mechanical industry.

With that said, we did not abandon the hypothesis that small enterprises showed limited flexibility. State promotion of investment appeared advisable for sectors and production lines identified as highly favored by demand imbalances in case of redistribution.

Thirty years after conducting this research, I continue to regard the flexibility of enterprises in structurally heterogeneous economies as profoundly limited. The discussion about global value chains has not established an increased capacity of enterprises in the Global South to free themselves from exploitative conditions. Under other circumstances, they might accomplish this by scaling upstream and downstream connections and engaging in the production of finished goods (Parker et al., 2014, 181; Özatagan, 2011, 898; Bernhardt & Pollack, 2016, 1238).

If the market does not trigger diversification, then, in structurally heterogeneous economies, rents must be used for promoting not-yet-competitive production lines in order to diversify them. At this point, it seems appropriate to invoke my theory around the contradictions of the state classes (Elsenhans, 1996, 173–254).

State classes are linked to the emergence of rent as the—ultimately political—consequence of structural heterogeneity. In theory, the historic mission of the state classes could be to overcome structural heterogeneity by using rent for overcoming rent and establishing integrated capitalist economies. Instead, however, they seem more inclined toward self-privileging. Naturally, this serves to deepen the rent-based character of their societies.

6 Integrating Globalization into the Social-Liberal Concept

The state classes singularly failed to dynamize their societies based on the availability of rent. Their strategy of import-substituting industrialization was replaced by manufacturing production for exports. This option for export-oriented industrialization does not eliminate rent. It merely leads to a more rational use of rent. Without internal reforms, such a strategy is too weak to transform these societies into functioning capitalist economies (Elsenhans, 2001). Export-oriented industrialization is an intelligent means of using rent. It can be successful if it is combined with rent-based industrial policies, and with internal reform in favor of rising mass incomes.

Globalization is often accompanied by low wages. However, it is not generally based on real low wages, but rather on low costs of labor in international prices through low exchange rates. It is rooted in devaluation below purchasing power parity. Some commentators have claimed that wages in the newly export-oriented industrializing economies are about one 20th, or even one 50th, of wages in the industrial West. However, nobody on such wages could survive in a Western country. Across much of the Global South, labor is cheap, mainly because the price level in comparison to Western countries is low. In some cases, the lowest 40% of households still spend around 40 to 60% of their household budget on food consumption. Here, devaluation depends on food self-sufficiency. Otherwise, higher world market prices for food would not allow devaluation below purchasing parity.

It is not an accident that the successful export-oriented countries were, at least initially, self-sufficient in food. The Arab world and sub-Saharan Africa are conspicuously absent from the list of regions which specialize in export-oriented industrialization. In these places, the Green Revolution was not successful, and it did not lead to self-sufficiency in food.

Self-sufficiency in major mass consumption goods, including in food, is the economic basis for devaluation below purchasing power parity. Export-oriented industrialization is based on devaluation, mostly below purchasing power parity. Real wages are about 1/6th to 1/8th of wages in the industrial West (Lafay, 1996, 948; Strack et al., 1997). This is approximately the same difference which obtained during the 1880s between the industrializing Mumbai region and Britain (Morris, 1965, 22; Lokanathan, 1935, 358).

Devaluation below purchasing power parity constitutes a mobilization of rent. Let us assume that a particular national workforce can buy more products locally than can be bought from its international currency counterpart of wages. This workforce must have access to wage goods in excess of what can be imported for its consumption from abroad. This excess must be produced locally, ultimately in the agricultural sector and the local small-scale industry. These provide the surplus which is bought by our workforce, which in turn receives an income in local currency. For this currency, there exists no counterpart in international sales.

Even for the most successful globalizers of the Global South, export shares never reached levels comparable to those of the leading industrialized countries of the West. Korea did not exceed 24% exports in GDP during the period where there was not yet scarcity of labor. Certainly, higher export shares followed the overcoming of marginality. However, they were largely absent during the period of overcoming marginality. Exports of goods and services in percent of GDP, for South Korea as (%) are to be found in World Bank, World Development Indicators (2022) where 1965 was 7.1, 1970 was 11.4, 1975 as 22.7, 1980 was 28.4, 1982 was 27.4, 1984 was 28.9, 1986 was 32, 1988 was 32.5, 1990 was 25, and 1995 was 25.7.

Certainly, export-oriented industrialization has a positive impact on local labor absorption. However, it would be unwise to consider that, for the industrialized world, the absorption of global marginal labor is possible only through exports. Tausch rightly insists on the existence of huge masses of poor and ultimately marginal labor in the Global South (40%, Tausch, 1978, 54).

Everything depends on the multiplier effects of additional exports. If we assume that there is structural heterogeneity, then competitiveness in a particular phase of the production process does not imply competitiveness in linked production, either upstream or downstream. Certainly, there exists an extensive literature on the decomposition of production lines in the value chain of production. But this is merely the expression of the limited spillover of competitiveness in a particular phase of production to linked upstream or downstream production.

The discussion has failed to mention two important mechanisms for overcoming these limitations. First, an egalitarian distribution of income increases the probability that the internal market will be extended in the wake of relatively small increases in export-oriented employment. Secondly, an intelligent policy of subsidizing linked upstream or downstream activities can broaden the effect of export orientation in manufacturing (Sid-Ahmed, 1990, 61).

The larger the market for simple industrial products, the likelier is the expansion of the productive apparatus beyond export-oriented production. An egalitarian agrarian reform creates a minimum income in agriculture. Certainly, this may be low. However, it affords labor some negotiating power in moving to export-oriented branches, as shown by China, despite its myriad contradictions.

Egalitarian agrarian reforms constitute the appropriation of rent incomes by average sized farmers. Such reforms allow these farmers to survive while mobilizing labor, which adds less to production than its share in total labor time. There is also a mobilization of marginal labor which would otherwise have been shed (Elsenhans, 1979, 552–564).

Some export-oriented branches are highly competitive on the world market. Burdening them with additional costs constitutes a mobilization of rent. This can be used for subsidizing diversification. One classic example are the taxes which were imposed on labor-intensive products made in Singapore. These taxes financed an upgrading of workers' skills. This facilitated a later shift to more skill intensive export production (Holtgrave, 1987, 74ff.).

There is a further means of mobilizing rents from the internal economy and from exports. This comprises an extension of the internal market and climbing the upstream and downstream ladders through industrial policy. This mobilization of rent can be rendered more efficient. The market must be allowed to do its work. Moreover, the government must intervene where the market is not yet producing the result of mass orientation of production and the localization of equipment production. This allows for the Kaleckian model to function.

Opting for such a combination of free market and state planning is characteristic of the social-liberal model advocated by Tausch (1989, 158, 2021b). The market can make important contributions for development. However, it must be embedded into a structure in which the demand for labor or other social structural conditions serves to empower labor. In this way, it is possible to establish the mechanism of mass production for mass consumption with increasing local investment goods production.

This requires a mental revolution, one in line with Tausch's conceptualization of the 1920s Austrian social democracy (Tausch & Prager, 1993, 3 f.; Tausch, 1993, 35). Deterministic political organization of production according to a strict economic plan carries obvious dangers. Thus, the market should be allowed to function where it can show positive results, and these positive results should be fostered by the mode of embedding the market into a global social structure. This structure should be the responsibility of the democratically legitimized state, which can operate it without derailing economic efficiency. Reformism is the basis of development policy. This is contrary to both orthodox Marxism and orthodox neoliberalism (Tausch, 1991, 133; Tausch, 2021a, 48).

References

Bernhardt, T., & Pollack, R. (2016). Economic and social upgrading dynamics in global manufacturing value chains. A comparative analysis. *Environment and Planning, 48*(7), 1220–1243.

Boatler, R. W. (1975). Trade theory predictions and the growth of Mexico's manufactured exports. *Economic Development and Cultural Change, 23*(4), 491–506.

Córdova, A. (1973). *Strukturelle Heterogenität und wirtschaftliches Wachstum*. Suhrkamp.

D'Adda, C., & Scorcu, A. E. (2003). On the time stability of the output-capital ratio. *Economic Modelling, 20*(6), 1175–1189.

Destanne, B. G. (1966). Industries industrialisantes et contenu d'une politique d'intégration régionale. *Economie appliquée, 19*(3/4), 415–473.

Destanne, B. G., & Ghozali, S. A. (1969). Les hydrocarbures et l'industrialisation de l'Algérie. *Revue algérienne des sciences juridiques, économiques et politiques, 6*(1), 256–294.
Elsenhans, H. (1977). *Algerien. Koloniale und postkoloniale Reformpolitik*. Institut für Afrikakunde im Verbund der Stiftung Deutsches Übersee-Institut.
Elsenhans, H. (1979). Agrarverfassung, Akkumulationsprozeß, Demokratisierung. In H. Elsenhans (Ed.), *Agrarreform in der Dritten Welt*. Campus Books.
Elsenhans, H. (1996). *State, class, and development*. South Asia Books.
Elsenhans, H. (2000). *La guerre d'Algérie 1954–1962. La transition d'une France à une autre. Le passage de la IV à la Ve République*. Publisud.
Elsenhans, H. (2001). Renten in der exportorientierten Industrialisierung. Zur notwendigen und widerspruchsvollen Rolle des Staats. In P. Gärtner (Ed.), *Staatlichkeit im Epochenbruch? Antworten aus der Perspektive des Südens und Ostens*. Hamburgisches Welt-Wirtschafts-Archiv.
Elsenhans, H. (2004). Overcoming rent by using rent. The challenge of development. *Intervention, 1*(1), 87–115.
Elsenhans, H. (2015). *Saving capitalism from the capitalists. A contribution to global and historical Keynesianism*. Sage.
Elsenhans, H. (2022). *Capitalism, development, and the empowerment of labour*. Routledge.
Elsenhans, H., Kleiner, E., & Dreves, R. J. (2000). *Développement, équité et extension du marché des masses. Une autre alternative. Le cas algérien. L'enjeu des PME industrielles*. Publisud.
Helmstädter, E. (1969). *Der Kapitalkoeffizient. Eine kapitaltheoretische Untersuchung*. Gustav Fischer.
Holtgrave, W. (1987). *Industrialisierung in Singapur. Chancen und Risiken industrieorientierter Spezialisierung*. Campus.
Kalecki, M. (1942). A theory of profits. *Economic Journal, 52*(2), 258–267.
Kramer, H. (1967). Entwicklungstendenzen der nationalen Befreiungsbewegung im subsaharischen Afrika. *Zeitschrift für Geschichtswissenschaft, 15*(8), 138–145.
Lafay, G. (1996). Les origines internationales du chômage européen. *Revue d'économie politique, 10*(6), 943–963.
Leontief, W. (1951). Input-output economics. *Scientific American, 18*(4), 15–21.
Leontief, W. (1954). Domestic production and foreign trade. The American capital position re-examined. *Economia Internazionale, 7*(1), 9–38.
Lewis, W. A. (1978). *The evolution of the international economic order*. Princeton University Press.
Lokanathan, P. S. (1935). *Industrial organization in India*. Allen & Unwin.
Marung, S. (2017). A 'Leninian moment'? Soviet Africanists and the interpretation of the October Revolution, 1950s-1970s. *Journal für Entwicklungspolitik, 33*(3), 21–48.
Modèle d'industrialisation par la P.M.I. (1985). In: Centre National d'Etudes et d'Analyses pour La Planification (ed.): Programme d'activités du Ce.N.E.A.P.1985/86. : Ce.N.E.A.P, pp. 27–44.
Morris, M. D. (1965). *The emergence of an industrial labor force in India. A study of the Bombay Cotton Mills, 1854–1947*. University of California Press.
Nohlen, D., & Sturm, R. (1982). Über das Konzept der strukturellen Heterogenität. In D. Nohlen & F. Nuscheler (Eds.), *Handbuch der Dritten Welt (1). Unterentwicklung und Entwicklung – Theorien, Strategien, Indikatoren* (pp. 92–116). Hoffmann & Campe.
Özatagan, G. (2011). Shifts in value chain governance and upgrading in the European periphery of automotive production. Evidence from Bursa, Turkey. *Environment and Planning, 43*(4), 85–90.
Parker, R., Cox, S., & Thompson, P. (2014). How technological change affects power relations in global markets. Remote developers in the console and mobile games industry. *Environment and Planning, 46*(2), 168–185.
Sid-Ahmed, A. (1990). *Development and resource-based industry. The case of the petroleum economies. A brief survey of recent studies*. OPEC Fund for International Development.

Singer, H. W. (1950). U.S. foreign investment in underdeveloped areas. The distribution of gains between investing and borrowing countries. *American Economic Review, 40*(2), 473–485.

Strack, D., Helmschrott, H., & Schönherr, S. (1997). Internationale Einkommensvergleiche auf der Basis von Kaufkraftparitäten. Das Gefälle zwischen Industrie- und Entwicklungsländern verringert sich. *IFO-Schnelldienst, 50*(1), 7–14.

Tausch, A. (1978). Politische Systeme und Abhängigkeit. Ein faktoranalytisches Modell. *Internationale Entwicklung, 9*(2), 54–62.

Tausch, A. (1987). *Jenseits der Weltgesellschaftstheorien. Sozialtransformationen und der Paradigmenwechsel in der Entwicklungsforschung*. Wilhelm Fink.

Tausch, A. (1989). Stable third world democracy and the European model. A quantitative essay. In B. Zbigniew & B. Hettne (Eds.), *Crisis in development. United Nations University European Perspectives Project 1986–87* (pp. 131–160). United Nations University.

Tausch, A. (1991). *Rußlands Tretmühle. Kapitalistisches Weltsystem, lange Zyklen und die neue Instabilität im Osten*. Eberhard.

Tausch, A. (1993). *Produktivkraft soziale Gerechtigkeit. Europa und die Lektionen des pazifischen Modells*. Eberhard.

Tausch, A. (2003). Social cohesion, sustainable development and Turkey's accession to the European Union. Implications from a global model. *Alternatives. Turkish Journal of International Relations., 2*(1), 1–41.

Tausch, A. (2011). Is globalization a driver of economic growth and social development? A cross-national, quantitative perspective on current, pro-globalist strategies of the European Union. *Society and Economy, 33*(3), 505–524.

Tausch, A. (2014). *The hallmarks of crisis. A new center-periphery perspective on long cycles. MPRA Paper 48356*. Munich Personal RePEc Archive.

Tausch, A. (2016a). Is globalization really good for public health? *International Journal of Health Planning and Management, 31*(4), 511–536.

Tausch, A. (2016b). Kaname Akamatsu. Biography and long cycles theory. In L. G. Efimovich, T. C. Devezas, & A. V. Korotayev (Eds.), *Kondratieff waves. Cycles, crises, and forecasts* (pp. 65–81). Uchitel Publishing House.

Tausch, A. (2018). Globalisation and development. The relevance of classical 'dependency' theory for the world today. *International Social Science Journal, 22*(4), 77–99.

Tausch, A. (2021a). *Smart development: The political economy in a post-COVID-19 world*. Nova Science Publishers.

Tausch, A. (2021b). *The socio-liberal society: Theories, perspectives, threats. An intellectual autobiography*. Nova Science Publishers.

Tausch, A., & Heshmati, A. (2011). Re-orient? Understanding contemporary shifts in the global political economy. *Journal of Globalization Studies, 2*(2), 89–128.

Tausch, A., & Heshmati, A. (2012a). *Globalization, the human condition and sustainable development in the twenty-first century*. Anthem Press.

Tausch, A., & Heshmati, A. (2012b). Migration, openness, and the global preconditions of 'smart development'. *Bogaziçi Journal Review of Social, Economic and Administrative Studies, 26*(2), 27–89.

Tausch, A., & Heshmati, A. (2013). Worker remittances and the preconditions of 'smart development'. *Society and Economy, 35*(1), 25–50.

Tausch, A., Grinin, L., & Korotayev, A. (2017). Akamatsu waves. In W. Thompson (Ed.), *Oxford research encyclopedia of empirical international relations theory* (pp. x1–x12). Oxford University Press.

Tausch, A., & Jourdon, P. (2011). *Trois essais pour une économie politique du 21e siècle. Mondialisation, gouvernance mondiale, marginalisation*. Harmattan.

Tausch, A., & Prager, F. (1993). *Towards a socio-liberal theory of world development*. Macmillan.

Timmer, M. P., de Vries, G. J., & de Vries, K. (2015). Patterns of structural change in developing countries. In J. Weiss & M. Tribe (Eds.), *Routledge handbook of industry and development* (pp. 65–83). Routledge.

von Bortkiewicz, L. (1907). Wertrechnung und Preisrechnung im Marxschen System (3). *Archiv für Sozialwissenschaft und Sozialpolitik, 25*(2), 445–489.

World Bank. (2022). World development indicators. https://datatopics.worldbank.org/worlddevelopment-indicators. Accessed 16 Mar 2022.

Analyzing Inequality, Dependence, and Poverty in the World System

Kunibert Raffer

Abstract Arno Tausch is one of the extremely few pioneers of development studies in Austria. This essay analyzes his work as a *dependencia*, showing that he still sees and uses dependency theory as a valuable tool of analysis, also in Eastern Europe and the EU. He is noted as one of the extremely few academics still showing the importance of relations of dependence. Finally, his relation to the idea of Unequal Exchange is explored, a concept he has never explored theoretically, but often referred to in his writings.

1 Introduction

Arno Tausch is one of the extremely few pioneers of development studies in Austria. He chose to work on Third World problems at a time when institutionalized development studies simply did not exist in academia. In those days it was pure chance whether a scientist chose to focus on development problems—which seldom happened—or not. Thus, a sociologist choosing to do research on development issues could more easily have chosen to do research on, say, aged people. A historian was much more likely to focus on European history than on the South. Economists choosing development economics rather than engaging in, e.g., those mathematical economics that caused the shattering crash of 2008–10 or at least in Austrian economic policy were the absolute exception.

Needless to say, choosing the focus development issues did not really enhance one's academic career in those early days before the dawn of development studiers in Austria. Until this day this specialization is not helpful for an academic career in Austria. Critical minds might remark not totally without justification that this dawn has not been able to change into full daylight. Political science was relatively more open to development studies, as the number of colleagues engaged in Third World

K. Raffer (✉)
University of Vienna, Vienna, Austria
e-mail: kunibert.raffer@univie.ac.at

G. Segell (ed.), *Development, Globalization, Global Values, and Security*,
https://doi.org/10.1007/978-3-031-24513-8_3

studies proved. Arno Tausch was among those pioneers in all disciplines, who chose to work on (under)development.

Especially in his early years, Tausch focused his research on *dependencia* (dependency theory), inequality, and (under)development within North-South relations. Long cyclical fluctuations of the world economy have always been one of the issues that captivated his attention

As a political scientist he has naturally connected those to political evolutions, also discussing political economic aspects. He has always worked quantitatively, seeing political science as a rigorous quantitative discipline testing theories and hypotheses with statistical methods, such as factor analysis and correlations. Tausch's works therefore usually contain a large section of empirical results and data.

Nevertheless, Tausch (1979, 56) already warned that "factor analysis can never be better than the data, on which it is based." Just feeding data into computers in the hope that a theoretical framework might emerge via factor analysis "is a disastrous error" (ibid.). He clearly defined factor analysis as "a means" (ibid.). As Tausch's publications stand out, containing numerous statistical results it seems important to clarify that he has never been unaware of the limits and dangers of number crunching.

While he has always stuck to quantitative analysis—which made him distinctive within political science—his focus shifted later. After the fall of the Berlin Wall, he turned his attention to Eastern Europe, as quite a few international development scientists did. When turning to Eastern Europe, Tausch also applied the concept of dependence and inequality to the relations between East and West, such as in Tausch (1985). During his nearly 15 years as an Austrian public servant, he turned his interest to the European Union, its global role, including its enlargement. There too, he focused on inequality, dependency, and development.

2 Tausch the Pioneer: Early Studies in North-South Relations

Tausch discussed *dependencia* early on, probably the first to do so in Austria, well before the first conference on development issues in the country, whose conference volume was edited by Raffer (1979). This conference was initiated by the Austrian Students' Association and subsidized by the then Minister of Science and Research, Hertha Firnberg. The association of Austrian development researchers was founded at Mattersburg in 1981, where all academics working on development gathered for the first time, deriving its name—Mattersburger Kreis (Mattersburg Circle)—from this small town. One of the participants was Arno Tausch, thus also a co-founder of the Kreis. It founded the Journal für Entwicklungspolitik—Austrian Journal for Development Studies by its English name—Austria's still existing academic development studies journal.

For the first time Firnberg introduced a special allotment of courses on development issues in addition to "normal" courses faculties would offer every term in the early 1980s. This helped development studies enormously, because no other courses had to be canceled in order to teach these, now additional, courses on development and underdevelopment. Faculties were quite happy about this solution. Development studies—nowadays still not as important as in many other countries—took off in Austria quite a bit after Tausch's article of 1973. Calling him a pioneer of Austrian development studies is thus wholly justified.

In the same year, the conference volume by Raffer (1979) was published, Tausch (1979) published his book on poverty and dependence, discussing political and economic problems within peripheral capitalism. *Dependencia* originated in Latin America, but gained supporters elsewhere too, such as most notably Samir Amin. Remarkable for such a decidedly heterodox approach, it even was presented briefly in the "temple" of neoclassical orthodoxy, the American Economic Review (Dos Santos, 1970).

Naturally, ruling orthodox theory soon "forgot" about this approach, as orthodoxy has successfully "forgotten" other irrefutable or embarrassing arguments against its teachings, most notably the Graham Paradoxon that proves that specialization according to comparative advantages does not necessarily lead to improvements for all, but is more likely to destroy one economy (cf. Raffer, 1994). Too publicly known, the Prebisch–Singer thesis could not be dealt with in this way. A *condemnatio memoriae*, as so aptly described in Blair's famous novel 1984, was impossible. Thus, Prebisch–Singer was successfully "re-interpreted" as a purely statistical debate (cf. ibid.).

Tausch (1979) is a "revised version of work done in my PhD-project" (ibid., 11), although substantial empirical parts had to be eliminated due to space restrictions. Nevertheless, he can present results of factor analysis as well, and the book has a large empirical part. After a brief introduction of *dependencia* thinking (considering the volume of the book), he discusses dependence in Africa, the political economy of apartheid, presents the immiseration by the Brazilian model of development as an example of not successful industrialization in the Third World. He interprets all these evolutions within the concept of the world system. Tausch explains the intention of his book as operating a "negative transfer of technology" (ibid.), meaning presenting ideas of colleagues in the Third World, especially in Latin America, to European colleagues. In fact, theories from the South have normally not been paid attention to, especially so in Austria.

While this is a general shortcoming of Northern science—possibly a remnant of colonialism—economists are in particular guilty of brushing away arguments from the South, arguably not just because they originate there (for too many "scientists" already a sufficient reason, though), but also as they disturb the cozy orthodox picture of the benevolent market that has been preached by orthodoxy, a way of thinking that lost the right to be called a science long ago. Naturally, history is also banned from orthodox theory. Colonialism and the distortions done to Southern economies are simply suppressed, as facts would destroy the assumption that all countries started from the same level of underdevelopment and some countries not hindered by non-European values have just been more adroit (for a very brief economic history disproving orthodoxy, cf. Raffer, 1987, 134–158). It appears that political science is quite similar. In both sciences exceptions confirming the rule exist.

3 Tausch, the *Dependista*

Dependency is one of the topics Tausch has always continued to research. Like Unequal Exchange it is one of the roots of inequality. As late as 2018, he published on the relevance of classical dependency theory in the new area of globalization (Tausch, 2018), a paper reprinted from the same journal where it had been initially published in 2010. The concept of some economies depending on others has always been present when he analyzed global structures.

According to Tausch this article aims at showing "the basic validity" of *dependencia* nowadays. Tausch (1998, 79) poses the rhetorical question whether "we in the global 'North'" should not finally learn "the lessons offered" by the periphery. He argues that Europe has joined the set of dependent regions, now "that dependency and penetration by multinational corporations have risen in Europe to staggering proportions," (ibid.):

> Thus, the early formulations of quantitative dependency, written in the late 1970s, gain some renewed relevance today, because the phase of the long economic cycle, which we currently face, is like the phase of the economic cycle in the 1970s. The only difference with the 1960s and 1970s is that the Europeans have nowadays become a heavily MNC [multinational corporation]-penetrated semi-periphery. (ibid., 95)

Reacting to discussions in Germany, Tausch (1986) asked the question what has remained of *dependencia* and world system analysis. His brief conclusion: "most notably the . . .interest in social transformations as preconditions of successful development." This quite general résumé seems to fit any social science and is surprisingly little. He interprets the history of Taiwan and South Korea as confirming "that accidents of world history have caused reforms." Thus, Tausch concludes, their development success was not due to authoritarian governments.

Tausch is right referring to the unique situation of both countries. Both divided between the two blocks, the "Western" parts enjoyed strong support by the USA in order to show economic success and to reinforce these "bulwarks" against Communism. They were allowed economic policies that in other countries would have triggered strong US reactions, most likely military interventions, or CIA coordinated putsches such as in Chile against Allende.

Surprisingly, Tausch does not go further than denying that authoritarian regimes as such—also called development dictatorships—were the key to development. He could have pointed out that Taiwan and South Korea followed the advice of Raúl Prebisch (1949), H.W. Singer (1950), and dependency thinkers. Singer (1986, 6) emphasizes the strong influence of Prebisch's ideas (modestly omitting his own influence) in South Korea, pointing out that many Korean economists were very conscious of that heritage (cf. also Raffer, 2000). Economic history shows that both "Asian Tigers" did not follow orthodox advice, the advice to open to the world market. In contrast to Latin America, which is usually associated with dependency and Prebisch, economic policies recommended by these unorthodox approaches were rigorously followed through in East Asia.

Governments intervened strongly, opportunities to export were used to the full, but they opened their own economies slowly and selectively to the world market. Both imports and foreign direct investment remained heavily restricted, domestic industries enjoyed heavy protection. Simultaneously exports were massively subsidized and supported (for details v., e.g., Raffer & Singer, 2001, 138–158). Taiwan and South Korea selectively dissociated from the world market. So strongly so that Wade (1990) coined the phrase "'Governing the Market," which characterizes East Asian economic policy in a nutshell. Clearly, such policies would have led to sudden upheavals of which the CIA would have been "innocent" or to military invasions in any other part of the world. In any case all newly "democratically installed" post-invasion authoritarian dictatorships would have opened the country to the world market. History, indeed, opened the way for selective dissociation.

Tausch is doubtlessly right. One must understand his remark on authoritarian regimes: they are no sufficient condition for development, as Latin America amply proves, and dependistas clearly showed. Unfortunately, a lot of critical work on these two dragons did not yet exist in 1986. So Tausch could not refer to them. Nevertheless, one would have expected more than this brief résumé in 1986, when financial dependence was very strong, and the Bretton Wood Institutions, as well as the Paris Club governed debtor countries in a neo-colonial way, though not the "dragons" yet. Inexplicably, dependency theory had virtually been totally rubbed out when increased dependence developed, some extremely few exceptions such as Tausch notwithstanding. Especially regarding financial dependence (in other words: debt crises officially acknowledged after 1982) had increased Southern dependence in a way *dependencia* theorists are quite likely to have considered impossible or at least unlikely during the heydays of *dependencia* some years earlier. With dependence increasing, dependency theory suddenly disappeared, a phenomenon that cannot be explained. While ruling academic theory has always enjoyed and been based on all too eager, uncritical followers—far too many of them to be academically good for science itself—such a total silencing of a once at least quite audible heterodox approach is unique.

One of the extremely few "survivors," Tausch continues to see dependency theory as a valuable, presently applicable tool of analysis, once again swimming against the (nowadays neoliberal) current. Tausch (2018, 82) speaks of a "rediscovery . . .of dependency theory." Stating that "The rise of 'critical economic theory' during the crisis of the 1930s (Keynes and Schumpeter), and the 1970s (dependency theory) is not a coincidence, and the time is ripe again for a thorough rethinking of economic theory, especially development theory during the current global economic crisis." (Tausch 2010, 83f) applies *dependencia* analysis to the European Union (EU). Measured by the indicator penetration by transnationals, the

EU "is one of the most penetrated continents [sic!] in terms of international capital penetration, worldwide" (ibid., 93).

He concludes that "Some of the predictions of world-system analysis and dependency theory can also be read today as an explanation for the unfortunately failed 'Lisbon strategy' of the EU-27" (ibid., 92). This strategy had aimed at making the EU the most competitive economy in the world. Anyone faintly familiar with how the EU works would have to consider the strategy of making this septuagenarian author the next world champion in ski jumping, dominating the whole field of jumpers, as relatively much more realistic, unless managed by the EU. However, what can one expect of a totally incompetent EU engaged in rolling back democracy, even unable to implement the public wish of abolishing the so-called daylight saving time over years, although that only means re-introducing what had been normal in Europe over centuries. In other words, one is not capable of closing again an open door that not long before had been closed. In the case of the EU dependency on the USA mingles with utter incompetence.

4 Tausch and Unequal Exchange

So, to say by necessity, Tausch also saw Unequal Exchange as an important mechanism of dependency. Focusing on dependency, inequal development, and poverty, one cannot avoid referring to it as the way by which value—be it in the Marxian sense or value added in neoclassical terms—is transferred from the South (the periphery) to the North (the center). This transfer—though not Unequal Exchange as presently defined yet—was already shown by Singer and Prebisch.

Clearly indicating inequality of remuneration of homogenous factors of production the Prebisch–Singer thesis rocked the boat of professional complacency exposing an apparent contradiction between theoretical expectations and practical outcome. Secularly deteriorating Southern net-barter-terms-of-trade (time series on double factoral terms of trade were not yet available) as documented by the seminal work of Singer and Prebisch destroy the whole established logic based on a beneficial world market (cf. Raffer, 1986). As mentioned above, this result was cleverly reduced to a purely statistical debate. Its essential point—that the world market does not function as orthodoxy purports, quickly vanished from sight. Data on which the Thesis is based show that inequalizing foreign trade siphons income off to the North. Thus, the main point of Unequal Exchange was already documented by the Prebisch–Singer Thesis and suppressed by orthodoxy. Right or wrong, the market is always right for the orthodox community, an attitude that makes one doubt whether economics is a science rather than a fanatic religion. The Holy Inquisition could indeed lean from economic orthodoxy. If it had, people (some troublemakers excepted) would still "rightly" believe the flat earth to be the center of the universe.

Emmanuel (1972) qualified Prebisch–Singer as on the fringe of Unequal Exchange. This seems unduly stern, even wrong, as they were the first documenting inequality in exchange, despite the non-existence of appropriate statistical data at that time. The generally accepted definition of Unequal Exchange, double factoral terms of trade different from 1 as formulated by Amin (1976, 168), covers in a way also the results of Singer and Prebisch.

Unequal Exchange became one of the hotly discussed topics in development studies, and within the *dependencia* school, after Emmanuel's (1972) book, which still based its conclusions on the Marxian transformation of values into prices. It was taken up and developed further by academics such as Amin (for a survey of various theories of Unequal Exchange v. Brolin, 2006; Raffer, 1987). Raffer presented his own "neoclassical" model independent of the Marxian value scheme comparing real income globally. Unequal remuneration in real terms means Unequal Exchange. Neoclassical market demands that equal factors be paid equal real remuneration as a precondition for its functioning. As his arguments could not be invalidated, nor brushed away as Marxian, it was unsurprisingly also simply "forgotten." It is little surprise that Tausch, always standing up for international justice and equality has also defended the existence of Unequal Exchange.

Tausch has never presented his own model of Unequal Exchange, nor participated in theoretical discussions on definitions, but often referred to Unequal Exchange. Quoting Kalecki, Tausch (2018, 83) asserts that "the share of wages in the capitalist center can only remain stable during rising monopoly power due to the rising 'unequal' exchange relationships." Always academically interested in long waves, he especially highlights "The affinity of dependency research and 'Kondratiev cycle research'" (ibid.).

Tausch referred in various publications to the issue of Unequal Exchange, quite logical if one considers his research. But he has no own definition of Unequal Exchange (at least to this author's knowledge) and has usually not referred to any concrete definition. It seems that he adopted Köhler's. Despite his not clearly defining what he meant by Unequal Exchange, one must suppose that he subscribes to Köhler's (Köhler & Tausch, 2002, xi, 61). Although the book's introduction is not signed, we must assume that both authors wrote it and agreed on the definition. Köhler (ibid., xi, 61) defines Unequal Exchange as

$$T = d*X - X \quad (1)$$

T is "the transfer value" gain or loss by Unequal Exchange
d is the "exchange rate deviation" or ERDI (Exchange Rate Deviation index)"

Quite problematic, *X* is "the volume of exports from a low- or middle-income country to the high-income countries" (ibid., 61). As already pointed out by Raffer (2006) this does not define Unequal Exchange, even though it might show certain inequalities. *X* (per defintionem prices x quantities) may be exchanged for whatever exchange rate. Naturally, strong deviations of purchasing power parity are likely to indicate inequality of remuneration. The real inequality of exchange is people doing the same task being rewarded differently in real terms (here PPP may be brought in). Or a worker carrying a bag of coffee on board in Africa is paid less (in real terms) than a worker carrying the same bag off the ship in Europe. This may occasionally—

by happenstance really—but not normally be caught by Definition 1. Export values corrected by whatever exchange rate or by referring to PPP do not automatically represent Unequal Exchange.

As *X* is prices x quantities, higher quantities may make up for lower remuneration of factors of production. More exploitation may make exchange according to Köhler's definition less inequal. Köhler equality may mean highly Unequal Exchange if quantities made up for miserable remunerations in real terms.

However, Tausch's apparently accepting a wrong definition of Unequal Exchange does not invalidate his results because he has never in his own work expressed how Unequal Exchange should be defined; Köhler and Tausch (2002, xi) arguably being the exception confirming the rule. As inequality in exchange demonstrably exists, Tausch's references to Unequal Exchange—without clearly specifying a particular definition or model—are logically correct.

5 Conclusion

Arno Tausch is one of those few heterodox scientists who have always been prepared to stand up for what they believe, to defend his research results, and to swim against the current if his research results suggested explanations differing from the mainstream. He has always worked interdisciplinary, not just within the social sciences, but also reaching out to wholly different disciplines. He has, e.g., included divinity. This can be explained by his interest in Liberation theology, which in turn includes economic, political, and sociological issues. He demanded: "The time has come to seriously reflect again on the common ground among Divinity studies and the social sciences, which can only consist in a renewed impetus for reform with and in the service of the poor, who are the great majority on our globe, regardless of their faiths and their cultures" (Tausch, 2000, 151). This is not his only reference to Liberation theology (cf., e.g., Tausch, 1986, 84)

After the de facto demise of dependency theory, he continued to apply it also to countries of the center and to Eastern Europe, clearly seeing its continuing if not increased importance for the world system. Thus, Arno Tausch is both a pioneer and a person forewarning against future problems.

References

Amin, S. (1976). *Unequal development, an essay in the social formations of peripheral capitalism*. Harvester Press.

Brolin, J. (2006). *The bias of the world, theories of unequal exchange in history*. Lund University Press.

Dos Santos, T. (1970). The structure of dependence. *American Economic Review, Papers & Proceedings., 60*(2), 231–236.

Emmanuel, A. (1972). *Unequal exchange, a study in the imperialism of trade*. Monthly Review Press.
Köhler, G., & Tausch, A. (2002). *Global Keynesianism, unequal exchange and global exploitation*. Nova Science.
Prebisch, R. (1949). El desarrollo económico de la América latina y algunos de sus principales problemas. *El Trimestre Económico, XVI*(3), 347–343.
Raffer, K. (1979). *Probleme des Nord-Süd Verhältnisses*. öH Press.
Raffer, K. (1986). Siphoning-off resources from the periphery: The relevance of Raúl Prebisch's thinking for the eighties. In H. W. Singer, N. Hatti, & R. Tandon (Eds.), *(1991), New World Order Series, (9): Aid and external financing in the 1990s* (pp. 583–606). Indus Publishing.
Raffer, K. (1987). *Unequal exchange and the evolution of the world system, reconsidering the impact of trade on north-south relations*. St. Martin's Press.
Raffer, K. (1994). Disadvantaging comparative advantages: The problem of decreasing returns. In R. Prendergast & F. Stewart (Eds.), *Market forces and world development* (pp. 75–90). St. Martin's Press.
Raffer, K. (2000). New forms of dependence in the world system. In A. Müller, A. Tausch, & P. M. Zulehner (Eds.), *Global capitalism, liberation theology, and the social sciences* (pp. 169–184). Nova Science Publications.
Raffer, K., & Singer, H. W. (2001). *The economic north-south divide: Six decades of unequal development*. Edward Elgar.
Raffer, K. (2006). Differences between inequalities and unequal exchange: Comments on the papers by Chaves and Köhler. *Entelequia, Revista Interdisciplinar, 2*, 197–199.
Singer, H. W. (1950). The distribution of gains between investing and borrowing countries. *American Economic Review, XL*(2), 478–485.
Singer, H. W. (1986). Raúl Prebisch and his advocacy of import substitution. *Development & South-South Cooperation, II*(3, Special Issue: Hommage to Raúl Prebisch), 1–12.
Tausch, A. (1973). Entwicklung der Unterentwicklung. Anmerkungen zur Dependenztheorie. *Lateinamerika Aspekte, 7*(3), 9–16.
Tausch, A. (1979). *Armut und Abhängigkeit, Politik und Ökonomie de peripheren Kapitalismus*. Braumüller.
Tausch, A. (1985). *Development, social justice, and dependence in Poland* (Vol. Occasional Paper No.9). Faculty of Economics, University of Sidney.
Tausch, A. (1986). Was bleibt von der Dependenztheorie? *Journal für Entwicklungspolitik, 3*(3), 84–90.
Tausch, A. (1998). *Globalization and European integration*. mimeo.
Tausch, A. (2000). Liberation theology and the social sciences: Seven hypotheses about the world capitalist system in our age. In A. Müller, A. Tausch, & P. M. Zulehner (Eds.), *Global capitalism, liberation theology, and the social sciences* (pp. 127–153). Nova Science Publications.
Tausch, A. (2010). Globalisation and development: The relevance of classical 'dependency' theory for the world today. *International Social Science Journal*, 467–488; reprinted in International Social Science Journal, 2018, 79–99.
Tausch, A. (2018). Globalisation and development: The relevance of classical "dependency" theory for the world today. *International Social Science Journal, 22*(4), 79–100.
Wade, R. (1990). *Governing the market*. Princeton University Press.

Politics Means Shaping the Shape

Raimund Dietz

Abstract Arno Tausch, together with his teacher Elsenhans, draws a political-economic vision that I fully support. Both say: Modern capitalism, insofar as it functions at all, does not function by itself, but only when it is shaped by a good and appropriate policy. As practical, realistic, and insofar true as this statement is, the social sciences in general find it difficult to support this statement with their methodology. This essay illustrates why this is so and, in all due brevity, outlines a conceptual framework for a political vision in the spirit of the author to be honored.

By political economics I understand a science that aims to develop sensible proposals for shaping the economy and society. The following terms are contained in the last sentence: politics, economy, society, shape (Gestalt), and shaping (Gestaltung).

What is it all about? From a purely *economic point of view*, it is about allocation of resources to cope with life under conditions of scarcity. This functional requirement, however, takes place within in a communicative *social context* of autonomously and spontaneously acting agents, whereby an unplanned "order" emerges. This process addresses the *social dimension* of the economy, that is the economy as a system.[1] The system has its own reality and effectiveness. The task of *politics* is to shape this spontaneous and autonomous order in such a way that it benefits society as a whole and that it can therefore function and develop sustainably.

The methodological shortcomings of the social sciences in general and political economy in particular lie in the fact that theory building focuses mainly on individual actions and neglects the systemic dimension. These shortcomings derive from the economic mainstream, whose theoretical core is a "logic of choice," or the figure of the homo oeconomicus. In this logic, the needy individual is confronted with

[1] For this, see Dietz (2018).

R. Dietz (✉)
Free University of Berlin, Berlin, Germany
e-mail: raimund.dietz@kabsi.at

G. Segell (ed.), *Development, Globalization, Global Values, and Security*,
https://doi.org/10.1007/978-3-031-24513-8_4

goods that he or she arranges in a utility-maximizing way. In other words, mainstream theory reduces its "architecture" to subject–object relations. Systems, however, cannot be derived from subject–object relations only. One can barely overestimate how much this reductionist methodology determines thinking in economics. It determines even macroeconomic approaches that implicitly operate with aggregates or with the construct of an average individual (Kirman, 1987; Wagner, 2020), thereby abstracting from intersubjective relations, even from exchange and money, through which society emerges and reproduces itself over time. One of the consequences of the reduction of economic theory to the instrumental dimension is that economics lacks the concept of Gestalt (shape).[2]

As a result, political theory fluctuates between two extremes: either believing that politics is superfluous or overestimating its tasks and believing that it is responsible for everything. Instead, political theory should find a middle ground, and this middle ground depends on the notion of a shape (Gestalt) that is to be shaped. But where does the concept of a shape come from? This is only possible if one overcomes the instrumentalism of mainstream economics and expands the architecture of the theory by including the systemic dimension.

We can only succeed in this if we introduce explicitly exchange into economics. Exchange relations require the medium of money which in its turns allows individuals to deepen their division of labor. The development of *money* is also paralleled by the development of the *State* and other supra-personal entities (Simmel, 1900/1978), leading in the last resort to what we call a civil society, that is a constantly renewing communicative context that has a *liberal order*. The comparison with language is obvious: during human development, communications produce the language that enables people to communicate with each other in a very differentiated way and to deal with their problems in a more civilized way than before. The emergence of civil societies is anything but natural.

Instead of Gestalt—an admittedly difficult term—the term "internal structure" is often used. "Internal economic and social structures are decisive for the impact of economic and social openness. It is vital that these internal structures are shaped according to social-liberal development principles." (Elsenhans, 2012, 1)—Structure, however, has a double meaning: it can be *material*, such as the structure of production (according to its origin, its use, etc.) or it can be *morphic*. Decisions ("rational choice") generate *material structures*, exchange (catallactic) processes, and generate *morphic structures*. A morphic structure is a kind of form-giving field, as we find them everywhere in nature and in society. Every "child" is capable of recognizing shapes. But the sciences struggle with them. Capitalism is such a "field." All great theorists, Marx, Weber, Keynes, Sombart, etc. worked explicitly or implicitly with the Gestalt concept. Marx was the most explicit on it. However, he strives for overcoming the capitalist morphic order by a non-working utopia. Weber tries to capture this "field" under the term "spirit of capitalism." Keynes states that the

[2] Classical physics explains structures by external factors. Modern systems theory relies on the idea of endogeneity and self-organization. This is the only way to derive the concept of Gestalt.

CLASSICS, under which he subsumes both the thinking of the classics and that of the neoclassics, is determined by a theoretical structure which offers no room for this Gestalt concept. And he clearly recognizes that the monetary economy functions according to principles different from those formulated by the CLASSICS in their models. Carried by this insight Keynes announced in 1933 that he wanted to break away from CLASSICAL principles and reconstruct economic theory as a theory of the money economy, that is to give his theory a new orientation. Unfortunately, he never fulfilled this claim. Throughout his life, he argued macroeconomically, i.e., with aggregates, and thus abstracted from the communicative and endogenous processes on which the liberal economic order of capitalist societies is based.

Without the concept of *Gestalt*, the rationale for political action hangs in the air. A "logic of choice," the teleological counterpart to the mechanistic causal concept of classical physics, suggests two possible extremes: Politics is to pursue the common good, or in contrast, is aiming only at coming to power, maintaining it and to exploit politics for its own ends. While the aforementioned goals know no "boundaries," the notion of "Gestalt" allows for a more down-to-earth concept of politics: it places on policymakers the task of shaping the "Gestalt," i.e., maintaining the system and adapting it to meet the challenges it faces permanently, both from the outside and from within. The two biggest claims on the system are the issue of equity and environmental sustainability. Policies are measured by whether they meet these goals.

In order to formulate the task of politics in this way, we need the Gestalt concept. As already said, this has been lost to economics, and Keynes also contributed to this by recognizing money not as a means of exchange but as a means of calculation, thus denying economic theory access to the concept of Gestalt. As a result, all Keynesians have the taboo of seeing money as a medium of exchange, which forces them to provide alternative justifications for money—an impossibility! From then on, for Keynesians money exerts all kinds of functions—except the role it actually plays. Sociologists and cultural theorists generally follow the Keynesian "narrative." But money is a medium of exchange! And because it is a medium of exchange, it is also a means of payment, a medium, a symbol, a means of calculation, a means of deferred payment, of capital accumulation, etc. Money is the center around which the economy revolves, and which virtually forces the economy to grow—while the mainstream declares money to be neutral!

But apart from this theoretical misconception, I agree with Keynes' position and share my concern with Arno Tausch's findings: the system of capitalist monetary economy depends on macroeconomic intervention, i.e. collective action, to keep it on track. Hayek, too, speaks of an order or Gestalt that arises spontaneously. But out of an unjustified belief in an automatism of equilibrium, he regards spontaneous order as sacrosanct. Consequently, he believes that the State should not intervene and should abstain from shaping the shape. Hayek is right in assuming that collective action cannot create order, but wrong in believing that order could survive without collective action.

References

Dietz, R. (2018). *Geld und Schuld - eine ökonomische Theorie der Gesellschaft, 6., durchgesehene Auflage*. Metropolis-Verlag.

Elsenhans, H. (2012). *Kapitalismus global. Aufstieg – Grenzen – Risiken*. Kohlhammer.

Kirman, A. (1987). *The intrinsic limits of modern economic theory: The emperor has no clothes.* EUI-Working Paper No. 87/323.

Simmel, G. (1900/1978). *Philosphy of money* (D. Frisby, Hrsg.). Routledge.

Wagner, R. E. (2020). *Macroeconomics as systems theory: Transcending the micro-macro dichotomy*. Palgrave Macmillan.

Multinational Firms and the Pattern and Structure of Foreign Direct Investment

Manfred Fuchs

Abstract A central building block of the prevailing development discourse, which was and remains deeply euro- and ethnocentric, remains the debate between modernization and dependency theories. Although my observations may appear to be a truncated debate from a historical perspective, this paper emphasizes that the fundamental issues that characterized this early discussion still form an essential core to influence current questions about the scope and meaning of development that remain unresolved. The promises of industrial modernization and progress stand on shaky grounds in the face of environmental destruction and decay. The unfortunate but widely accepted label of developing countries as "backward," as was standard in the 1960 and the 1970s, cemented a widely uncritical belief that development equates with industrial progress, which has proven to be a fatal delusion.

Arno Tausch during a political science seminar, distributed a reasonably long reading list with reference to the world systems theory. The year was 1985, and I distinctly remember the hardship of working through those fascinating but arduous readings. At that time, providing access to all journal articles was not easy; in contrast with today's online access to information, it was a different world. Arno was an energetic, emphatic, friendly, supportive, and articulate teacher who encouraged us to be critical and self-reflective and develop a perspective that would force us to look beyond the narrow confines of our disciplines. When I scroll through this extensive literature list, it is a summary of literature that, on the one hand, reveals the core, and on the other hand, presents the dead end of the development discourse in the 1970s and the early 1980s in a retrospective of almost forty years.

A cornerstone of this discourse, which remained profoundly euro- and ethnocentric, was the dispute between modernization and dependency theories on both sides (Frank, 1966; Cardoso, 1982; Prebisch, 1984; Rostow, 1987, 1997; Wallerstein, 2000, 2012; Maddison, 2001, 2003; Cardoso & Campodónico, 2017). Although my

M. Fuchs (✉)
University of Graz, Graz, Austria
e-mail: manfred.fuchs@uni-graz.at

G. Segell (ed.), *Development, Globalization, Global Values, and Security*,
https://doi.org/10.1007/978-3-031-24513-8_5

observations on this debate may appear to be somewhat truncated, I am of the view that these underlying issues, which characterized this early discussion, form a core and, by and large, it continues to affect the current debate on the scope and meaning of development, which remain unresolved. While in a perhaps unfortunate yet widely accepted formulation, the German sociologist Eric Weede described developing countries as "backward" (*rückständig* or underdeveloped) (Weede, 1985, 13), repeating a widely accepted assumption, and concluded further that especially in these particularly "backward countries," any progressing development "is initially escorted by the aggravation of an unequal income distribution" (ibid).

Andre Gunder Frank contrasted this perspective with his groundbreaking analysis of the "development of underdevelopment" published in 1966, at a time when modernization theory was experiencing its heyday (Frank, 1966, 1989). Around the same time, Raul Prebisch criticized how peripheral countries which were integrated into a prevalent postcolonial world economy in fact exploited their rich natural resources and raw materials by allowing first-world multinationals to exploit these resources (Prebisch, 1981, 1984; Clairmonte, 1987; Dunning, 1988; Cantwell et al., 2010). Thus, attracting foreign direct investment (FDI) did not mechanically generate a sound development, especially in resource-rich economies in the third world (Raczynski, 1989; Arnold, 2014; Southgate et al., 2019).

In the 1960s, Raul Prebisch identified parts of the problem: Western "technolog [ies] of the centers had penetrated mainly into activities connected with [these] primary exports [...] but not into other activities [...] where the productivity of a very large proportion of the labor force was very low," and unsurprisingly, he concluded that the "basic problem of development, therefore, involved raising the level of productivity of the entire labor force" (Prebisch, 1984, 177). While the level of productivity is one matter, and the cause and nature of poverty another, both are interdependent and raising the productivity does not automatically eradicate the latter. Prebisch's message was turning an old question around; that is, how it is possible to integrate developing countries more fully. However, he did not specify integrating them more equally in a world economy still dominated by a few industrialized economies. Seeking an answer that more integration, particularly pushed along export-oriented growth, has become a widely accepted economic policy to promote development (Weiss, 1991; Moon, 1998; Corbo, 2019).

Nevertheless, it was acknowledged that the benefits of this nature of growth did not easily trickle down to the "large masses of the low-income population" living in population-rich developing economies (Prebisch, 1984, 181). It is noteworthy that the term underdeveloped transmuted to developing within a decade in the 1950s, because it was assumed that the new, albeit poor nations, would catch up with the growth model practiced in the West. Walt W. Rostow, reflecting on the development process in that decade, conceded that his efforts of working on a development theory forced him "to mobilize all what he learned to answer to basically big questions, one how economic theory was applied to explain economic history" and likely the trickier question, "how economic forces interacted with the whole life of societies" (Rostow, 1984, 229).

In an essay, reflecting on the impact and consequences of his "Stages of Economic Growth" (Rostow, 1991), he stated that in a bipolar world, it is not surprising that the idea of development became a political agenda that emerged from two competing and struggling worldviews in a new geopolitical postcolonial and conflicting world. Rostow's stage model was designed as a dynamic theory of development, based on five interrelated but successive stages. Briefly, the story was the following: For a long time, traditional (agricultural) societies were limited in their capabilities to increase productivity (Rostow, 1991, 5).

A lack of technique that could enable them to break this low productivity was identified as the core limitation. This period was the first stage. The second stage, which was termed as the "pre-take off period," was characterized by its slowly evolving dynamism to transform all that was summarized in the preceding stage, as the major limitations of traditional societies. This slow but substantial transformation, according to Rostow, first occurred in Western Europe in the late seventeenth century. In that period, "new types of enterprising men" (Rostow, 1991, 6) emerged and in a "private economy" separated from a seemingly collective traditional economy, described by its characteristics to "mobilize savings [...] take risks in pursuit of profits" (Rostow, 1991, 7).

From that initial, magical ignition, the emerging transformation increasingly enabled an "investment in transport, communications, and in raw materials" (ibid) increasing the "scope of commerce" (ibid). Therefore, the whole question boiled down to how to get rid of the backwardness of traditional societies, commercialize traditional societies, make them accumulate savings, and implement a somewhat premature entrepreneurial spirit; all in all, how to get rid of the so-called existing backwardness?

After the second pre-take off stage, "[w]e come now to the great watershed in the life of modern societies" in which "the rate of effective investment and savings [...] rise" and "new techniques in agriculture" rapidly expanded (Rostow, 1991, 10). In this third stage, the final or full transformation of these traditional societies took place. In the take off stage, the economy was assumed to revolve around a "a relatively narrow [set] of industries and technologies" (Rostow, 1991), but in the mature stage, which automatically followed (the fifth stage), these new transformed societies diversified to "absorb and to apply efficiently over a very wide range of resources—if not the whole range—[and become capable of harvesting] the "fruits of [...] modern technology" (Rostow, 1991, 10).

To one's surprise, consequently, real income levels rose, consumption levels became increasingly higher, and the golden age of Fordism was spreading, yet not equally and not without destroying the sustainability of lives. Yet, the transformation from traditional to modern occurred (Dunford, 2009; Okuma & Harada, 2022). However, not everyone was working in these "offices or in skilled factory jobs [...] to acquire the consumption fruits of the mature economy" (Rostow, 1991, 10–11), but it remained a powerful myth, degrading all leftovers as subaltern parts, and all traditions as the histories of minorities (Chakrabarty, 2008).

The debate moved on, and the major problem essentially remained how developing countries would succeed in initially taking off and subsequently move along an almost inevitable process of modernization. Low or zero savings rates and lack of capital and technologies in poor economies became a focal issue in solving the puzzle of development in the 1980s and the 1990s (Bornschier, 1982; Tausch, 1993, 2018, 2019; Amadeo & Pero, 2000; Bair & Werner, 2017).

The question as to how development, and especially the expected Western growth model, could be exported and spread across the world without increasing the harsh dependency and enduring poverty of the developing economies persisted (Tausch, 1993). It is worth noting that it is a growth model still celebrated by many despite its social, political, and environmental costs (Milanovic, 2014; Piketty, 2014; Facchini & Couvreur, 2015; Milovanović, 2015).

All these ideas emerged by scrolling through Arno's seminar literature list and the comments made by him in Innsbruck in 1985. Yet, all these selective and short quotes provide only a glimpse of a wider and more profound impact caused by a still active worldview about the means and ends of economic growth.

Looking back by five decades, in the early 1960s, the underlying and almost paradigmatic division of labor between the north and south was slowly but significantly changing (Fröbel et al., 1978, 1980; Sturgeon et al., 2008; Ernst, 2009; Casson, 2013; McWilliam & Nielsen, 2020). However, this highly fragmented global division of labor of value-creating processes reveals the narrow scope of integration poor economies into a fast-growing and more fragmented world economy (Casson, 2013; Ernst, 2009; Fröbel et al., 1978). An issue that remained the source of important dispute was the question of the role that multinational corporations play in this fundamental upheaval (Hufbauer, 1976; Feenstra, 1998; Jones, 2003; Pitelis, 2011).

The fundamental issue about equal development is portrayed by the unequal growth of trade, fragmented and highly concentrated foreign direct investments measured by the inflows and outflows (in and from major regions), and unequal and prevalent pattern of per capita income; further, identifying the poverty line between economies is still a source of misery and conflict (Deaton & Dupriez, 2011; Bank et al., 2013).

What about more integration? How are poor countries integrated in the world economy? While this is not an easy question, it can be discussed using the data on world trade and FDI patterns. Currently, both sets of data are somewhat easily accessible (see for details WTO data and UNCTAD). A pivotal dimension of the global economy is international trade, which is broadly defined as the cross-border export and import of goods and services, signifying the scope and vitality of exchange among different nations. Trade data compiles the cross-border trade of goods (e.g., machines, automobiles, or energy) and services (Reinert et al., 2009; Reinert, 2021).

Trade has always been an integral part of civilization and can be traced back to the earliest stages of mankind (Tracy & Chaudhuri, 1991). Aggregated trade data tells a success story: Since 1948, the total merchandise trade has grown annually by approximately 8.7% (WTO, data 2021). But how is this trade distributed among regions and economies? In 2019, the European Union (EU) took the lion's share with approximately 37.7%, Asia's share increased (by more than double) from 14 to

34%, while North America (the USA and Canada) dropped from 21.6% to 9.0% (WTO data, 2021). Therefore, there is no question that the shares among regions have changed significantly (Fuchs, 2022).

It is important to examine these figures more closely. Approximately 50 economies account for 93% of all the total merchandise export and import trade and 33 economies for 80%. Thus, the integration (a factor debated by Prebisch in the 1960s) is highly unequal. In short, high income economies have a much greater intranet- and interregional share of trade, whereas poor economies in Africa, Latin America, and Asia exhibit only a level ranging from approximately 3 to 5% (WTO data 2021). However, these country-level data are still a bit misleading; the major actors in these interregional and intraregional trade are multinational firms.

The United Nations Conference on Trade and Development (UNCTAD) estimated that approximately 80% of trade is interfirm and intrafirm trade and shifting the focus to a small number of leading economies (only 20 countries hosting the largest and most powerful multinational firms) accounts for approximately 75% of the total merchandise trade (Beckerman, 1956; UNCTAD, 2021). Thus, although since 1948 global trade has expanded incredibly to 19 trillion USD in 2019 (WTO database, 2021), the unequal pattern remains, and the gap appears to be growing.

Does the pattern change much if we examine the data on FDI? Not by much. FDI is defined as an investment made in order to establish a "long-lasting relationship" reflecting the aim to regain or maintain the "control" over the investment taken (WIR 1996, 219). Control is exercised by large multinational firms. FDI was traditionally used to analyze the behavioral strategies employed by multinational firms (Razin, 2003). Firms involved in FDI seek to exploit firm-specific advantages created by its (widely assumed) superior efficiency gains or cost advantages (Mezias, 2002; Mudambi & Mudambi, 2002; Buckley & Ghauri, 2004; Buckley et al., 2007, 2016; Jung et al., 2008; Hayat, 2018; Duce & de Espana, 2021).

Yet, Stephen H. Hymer has noted that since the 1960s, FDI is the mechanism used by the multinational firms to exercise control and exploit their monopoly powers (Hymer, 1970, 1976). Thus, the FDI made by a multinational firm is not simply to diffuse advanced technologies and superior capabilities to and across foreign locations lacking capital or savings. Rather, these locations attract its natural resources and several other incentives with their cheap and disciplined labor force. Orthodox international business scholars defined four major motives for FDI—market-seeking, resource-seeking, efficiency-seeking, and strategic reasons (Dunning & Lundan, 2008)—but behind these motives, multinational firms exercise their superior market power and operational flexibility by exploiting their improved efficiency to organize cross-border value-adding activities (Kogut, 1985; Kogut & Zander, 1993; Verbeke, 2003).

Furthermore, how does the present-day pattern and structure of FDI look and how did it change from the past to today? As Peter Buckley and Richard Caves eloquently elaborated, in the 1970s, mostly multinational firms from a handful of industrial economies established subsidiaries in another few advanced industrial economies, which dominated the prevailing foreign direct investment pattern (Caves, 1974; Buckley & Casson, 2002). Using the data from several World Investment Reports compiled by the UNCTAD, approximately 90% of all FDI is allocated among major industry sectors operating in advanced industrialized economies (for a long time, it was in the USA, the UK, France, the Netherlands, and Germany). For example, in 1998, 71% of all FDI inflows moved into industrialized economies and approximately 25% was directed to developing economies (UNCTAD, 1999).

However, it is important to note that it was never a country, but a multinational firm that was investing in a foreign entity it was going to control; in 2010, almost 80% of all FDI was accounted for by mergers and acquisitions made in advanced economies (UNCTAD, 2011, 2012, 2013, 2021). In the early 1960s, most FDI went into mining and oil-extracting sectors in a resource-rich developing countries, but it must be acknowledged that this pattern changed substantially. In 2019, 52% of FDI (in and outflows) was accounted for by developed economies and 44% by developing countries.

However, the recipients remained a few major locations and fast-growing markets. In China, Hong Kong was the largest recipient, and Asia accounted for more than 57% of all FDI to the developing world; Africa accounted for less than 4%. However, four countries, Egypt, the Republic of Congo, South Africa, and Nigeria attracted the major share. Although FDI is generally acknowledged as a driver of economic growth (Streeten, 1972; Doh, 2019), the actual pattern of distribution within and among regions tells a different story. Whether growth is attracting FDI or FDI is generating growth is still a disputed and unresolved issue (Swan, 1956; Deardorff, 2001; Peng et al., 2011; Abbas et al., 2020; Sun et al., 2021).

All the questions that arise from discussing these trade structures and FDI figures revolve around the question of how these affect equality and development, those of which were prevalent in Arnos' literature list from 1985 and his subsequent works (Tausch, 1993, 2007, 2010, 2018, 2019). Some issues have been marginalized, and others outdated, for example, the critique of modernization theory and the analyses of dependency theorists. However, all these issues are still likely more relevant than ever, and the answer regarding how to support a more sustainable, peaceful, and equal world is becoming increasingly urgent.

These were the pressing questions that Arno eloquently discussed for a long time (Tausch, 1993, 2007, 2010). Beyond all the questions raised, the problem is how we change our behavior by becoming more sustainable, more peaceful, and leading a life contributing to a just society. However, William Easterly called growth a "precious object" that "either remains elusive or is a disappointment when found" (Easterly, 2002, X). In rich, affluent societies, it seems easy to blame growth and consumption patterns for environmental destruction and the waste of resources.

However, for millions of people living in abject poverty and suffering from disease, facing poor sanitation, lacking access to education, and confronted every day with the worry of putting food on the table, the blunt rejection of economic growth is anachronistic and cynical. The high mortality rate in many poor countries is caused by the prevalence of hunger—poor families, ethnic minorities, and women are the most vulnerable (Easterly, 2002, 12). Thus, income inequality, measured for example over decades, is only one part of the whole story. Reducing income inequality is important, but it will not suffice. Only equal opportunities, especially for the poorest, will allow all to live a decent and peaceful life without hunger and misery (Atkinson, 2015).

References

Abbas, H. S. M., Gillani, S., & Ullah, S. (2020). Nexus between governance and socioeconomic factors on public service fragility in Asian economies. *Social Science Quarterly, 101*, 1850–1868. https://doi.org/10.1111/ssqu.12855

Amadeo, E. J., & Pero, V. (2000). Adjustment, stabilization, and the structure of employment in Brazil. *Development Studies, 36*, 120–148. https://doi.org/10.1080/00220380008422640

Arnold, G. (2014). *The resources of the third world*. https://doi.org/10.4324/9781315061641.

Atkinson, A. B. (2015). *Inequality. What can be done?* Harvard University Press.

Bair, J., & Werner, M. (2017). New geographies of uneven development in global formation: Thinking with chase-Dunn. *Journal of World-systems, 23*, 604–619. https://doi.org/10.5195/jwsr.2017.727

Bank, W., Deaton, A. S., & Dupriez, O. (2013). *Measuring the real size of the world economy*. https://doi.org/10.1596/9780821397282_ch21.

Beckerman, W. (1956). Distance and the pattern of intra-European trade. *The Review of Economics and Statistics, 38*, 31. https://doi.org/10.2307/1925556

Bornschier, V. (1982). Dependence on foreign capital and economic growth: A reply to Weede and Tiefenbach's Critique. European Journal of Political Research 10, 445–450. doi:https://doi.org/10.1111/j.1475-6765.1982.tb00036.x.

Buckley, P. J., & Casson, M. (2002). *The future of the multinational enterprise*. 25th Anniversary Edition. Palgrave Macmillan.

Buckley, P. J., Chen, L., Clegg, L. J., & Voss, H. (2016). Experience and FDI risk-taking: A microfoundational reconceptualization. *Journal of International Management, 22*, 131–146. https://doi.org/10.1016/j.intman.2016.02.001

Buckley, P. J., Devinney, T. M., & Louviere, J. J. (2007). Do managers behave the way theory suggests? A choice-theoretic examination of foreign direct investment location decision-making. *Journal of International Business Studies, 38*, 1069–1094. https://doi.org/10.1057/palgrave.jibs.8400311

Buckley, P. J., & Ghauri, P. N. (2004). Globalisation, economic geography, and the strategy of multinational enterprises. *Journal of International Business Studies, 35*, 81–98. https://doi.org/10.1057/palgrave.jibs.8400076

Cantwell, J., Dunning, J. H., & Lundan, S. M. (2010). An evolutionary approach to understanding international business activity: The co-evolution of MNEs and the institutional environment. *Journal of International Business Studies, 41*, 567–586. https://doi.org/10.1057/jibs.2009.95

Cardoso, F. H. (1982). *Introduction to the sociology of "developing societies." 11*, 112–127. https://doi.org/10.1007/978-1-349-16847-7_11.

Cardoso, F. H., & Campodónico, H. (2017). The future of Latin America in the global economy. An interview with Fernando Henrique Cardoso. *International Development Policy, 16*, 22. https://doi.org/10.4000/poldev.2348

Casson, M. (2013). Economic analysis of international supply chains: An internalization perspective. *Journal of Supply Chain Management, 49*, 8–13. https://doi.org/10.1111/jscm.12009

Caves, R. E. (1974). Multinational firms, competition, and productivity in host-country markets. *Economica, 41*, 176. https://doi.org/10.2307/2553765

Chakrabarty, D. (2008). *Provincializing Europe. Postcolonial thought and historical difference*. Princeton University Press.

Clairmonte, F. F. (1987). Prebisch and UNCTAD – The unravelled myths. *Minerals Energy – Raw Mater Report, 5*, 14–33. https://doi.org/10.1080/14041048709409296

Corbo, V. (2019). *Export-oriented development strategies*. https://doi.org/10.4324/9780429038013.

Deardorff, A. V. (2001). Rich and poor countries in neoclassical trade and growth. *Economic Journal, 111*, 277–294. https://doi.org/10.1111/1468-0297.00608

Deaton, A., & Dupriez, O. (2011). Purchasing power parity exchange rates for the global poor. *American Journal of Applied Economics, 3*, 137–166. https://doi.org/10.1257/app.3.2.137

Doh, J. P. (2019). MNEs, FDI, inequality and growth. *Multinational Business Review, 27*, 217–220. https://doi.org/10.1108/mbr-09-2018-0062

Duce, M., & de Espana, B. (2021). *Definitions of Foreign Direct Investment (FDI) a methodological note*. Bank of International Settlements.

Dunford, M. (2009). The future of economic growth. As new becomes old. *Economic Geographer, 82*, 227–229. https://doi.org/10.1111/j.1944-8287.2006.tb00301.x

Dunning, J. H. (1988). *Explaining international production*. Routledge Revivals.

Dunning, J., & Lundan, S. M. (2008). *Multinational enterprises and the global economy*. Edward Elgar Publisher.

Easterly, W. (2002). *The elusive quest for growth*. MIT Press.

Ernst, D. (2009). A new geography of knowledge in the electronics industry? Asia's role in global innovation networks. *Electron Journal*. https://doi.org/10.2139/ssrn.2742923.

Facchini, F., & Couvreur, S. (2015). Inequality: The original economic sin of capitalism? An evaluation of Thomas Piketty's "Capital in the twenty-first century". *European Journal of Political Economy., 39*, 281–287. https://doi.org/10.1016/j.ejpoleco.2015.05.001

Feenstra, R. C. (1998). Integration of trade and disintegration of production in the global economy. *Journal of Economic Perspective, 12*, 31–50. https://doi.org/10.1257/jep.12.4.31

Frank, A. G. (1966). The development of underdevelopment. *Money Review, 18*, 17–31. https://doi.org/10.14452/mr-018-04-1966-08_3

Frank, A. G. (1989). The development of underdevelopment: A reprint. *Money Review, 41*, 37–51. https://doi.org/10.14452/mr-041-02-1989-06_4

Fröbel, F., Heinrichs, J., & Kreye, O. (1978). The world market for labor and the world market for industrial sites. *Journal of Economic Issues, 12*, 843–858. https://doi.org/10.1080/00213624.1978.11503581

Fröbel, F., Heinrichs, J., & Kreye, O. (1980). *The new international division of labour*. Cambridge University Press.

Fuchs, M. (2022). *International management. The process of internationalization and market entry strategies*. Springer Gabler.

Hayat, A. (2018). FDI and economic growth: the role of natural resources? *Journal of Economic Studies., 45*, 283–295. https://doi.org/10.1108/jes-05-2015-0082

Hufbauer, G. C. (1976). *International trade and finance*. https://doi.org/10.1017/cbo9780511895685.007.

Hymer, S. (1970). *The efficiency (contradictions) of multinational corporations*. https://doi.org/10.2307/1815843.

Hymer, S. H. (1976). *The international operations of national firms: A study of direct foreign investment*. MIT Press.

Jones, G. (2003). *Business history around the world*. https://doi.org/10.1017/cbo9780511512100.018.

Jung, J. C., Beamish, P. W., & Goerzen, A. (2008). FDI ownership strategy: A Japanese-US MNE comparison. *Management International Review, 48*, 491–524. https://doi.org/10.1007/s11575-008-0034-8

Kogut, B. (1985). Designing global strategies: comparative and competitive value-added chains. *Sloan Management Review, 26*(1), 15–28.

Kogut, B., & Zander, U. (1993). Knowledge of the firm and the evolutionary theory of the multinational corporation. *Journal of International Business Studies, 24*, 625–645. https://doi.org/10.1057/palgrave.jibs.8490248

Maddison, A. (2001). The world economy. *Development Centre Study*. https://doi.org/10.1787/9789264104143-en

McWilliam, S. E., & Nielsen, B. B. (2020). Global value chains and development: Redefining the contours of 21st century capitalism. *Journal of International Business Studies, 51*, 1347–1350. https://doi.org/10.1057/s41267-020-00303-3

Mezias, J. M. (2002). Identifying liabilities of foreignness and strategies to minimize their effects: The case of labor lawsuit judgments in the United States. *Strategic Management Journal., 23*, 229–244. https://doi.org/10.1002/smj.220

Milanovic, B. (2014). The return of "patrimonial capitalism": A review of Thomas Piketty's capital in the twenty-first century. *Voprosy Ekonomiki, 12*(1), 127–144. https://doi.org/10.32609/0042-8736-2014-9-127-144

Milovanović, M. (2015). Thomas Piketty's capitalism revisited. *Panoeconomicus, 62*, 663–677. https://doi.org/10.2298/pan1505663m

Moon, B. E. (1998). Exports, outward-oriented development, and economic growth. *Political Research Quarterly., 51*, 7–36. https://doi.org/10.1177/106591299805100101

Mudambi, R., & Mudambi, S. M. (2002). Diversification and market entry choices in the context of foreign direct investment. *International Business Review, 11*, 35–55. https://doi.org/10.1016/s0969-5931(01)00046-4

Okuma, K., & Harada, Y. (2022). Robert Boyer, Les capitalismes à l'épreuve de la pandémie, La Découverte, 2020. *Institutional Economic Review, 19*, 511–522. https://doi.org/10.1007/s40844-022-00235-7

Peng, M. W., Sun, S. L., & Blevins, D. P. (2011). The social responsibility of international business scholars. *Multinational Business Review, 19*, 106–119. https://doi.org/10.1108/15253831111149753

Piketty, T. (2014). *Capital in the twenty-first century*. The Belknap Press.

Pitelis, C. N. (2011). Globalization, development, and history in the work of Edith Penrose. *Business History Review, 85*, 65–84. https://doi.org/10.1017/s0007680511000031

Prebisch, R. (1981). Raul Prebisch on Latin American Development. *Popular Development Review, 7*, 563. https://doi.org/10.2307/1972587

Prebisch, R. (1984). Five stages in my thinking on development. In G. M. Meier & D. Seers (Eds.), *Pioneers in development* (pp. 175–192). Oxford University Press.

Raczynski, A. M. (1989). Mineral resources and Third World development. *Episodes, 12*, 255–256. https://doi.org/10.18814/epiiugs/1989/v12i4/004

Razin, A. (2003). FDI contribution to capital flows and investment in capacity. *Electron Journal, 23*, 23–32. https://doi.org/10.2139/ssrn.1009362

Reinert, K. A. (2021). *International economics. New perspectives on the world economy*. Cambridge University Press.

Reinert, K. A., Rajan, R. S., Glass, A. J., & Davis, L. S. (2009). *The Princeton encyclopedia of the world economy*. Princeton University Press.

Rostow, W. W. (1984). Development: The political economy of the Marshallian long period. In G. M. Meier & D. Seers (Eds.), *Pioneers in development* (pp. 229–261). Oxford University Press.

Rostow, W. W. (1987). *Employment and Growth: Issues for the 1980*. https://doi.org/10.1007/978-94-009-3605-8_4.

Rostow, W. W. (1991). *The stages of economic growth: A non-communist Manifesto*. Cambridge University Press.

Rostow, W. W. (1997). Lessons of the Plan: Looking forward to the next century. *Foreign Affairs, 76*, 205–218. https://doi.org/10.2307/20048111

Southgate, D., Disinger, J. F., & Norgaard, R. B. (2019). *Sustainable resource development in the third world*. https://doi.org/10.4324/9780429308079-8.

Streeten, P. (1972). *The Frontiers of Development Studies*. https://doi.org/10.1007/978-1-349-05017-8_6.

Sturgeon, T., Biesebroeck, J. V., & Gereffi, G. (2008). Value chains, networks, and clusters: Reframing the global automotive industry. *Journal of Economic Geography, 8*, 297–321. https://doi.org/10.1093/jeg/lbn007

Sun, C., Abbas, H. S. M., & Xu, X. (2021). Role of capital investment, investment risks, and globalization in economic growth. *International Journal of Finance Economics, 76*, 251–278. https://doi.org/10.1002/ijfe.2514

Swan, T. W. (1956). Economic growth and capital accumulation. *Economic Record, 32*, 334–361. https://doi.org/10.1111/j.1475-4932.1956.tb00434.x

Tausch, A. (1993). *Towards a socio-liberal theory of world development*. https://doi.org/10.1007/978-1-349-12282-0_8.

Tausch, A. (2007). Did recent trends in world society make multinational corporations penetration irrelevant? Looking back on Volker Bornschier's development theory in the light of recent evidence. *Electron Journal, 3*, 23–34. https://doi.org/10.2139/ssrn.977155

Tausch, A. (2010). Paul Boccara's analysis of global capitalism, the return of the bourbons, and the breakdown of the Brussels/Paris neo-liberal consensus: A quantitative analysis on the political economy of our time. *Electron Journal, 23*, 45–57. https://doi.org/10.2139/ssrn.1726316

Tausch, A. (2018). Globalisation and development: The relevance of classical "dependency" theory for the world today†. *International Social Science Journal, 68*, 79–99. https://doi.org/10.1111/issj.12190

Tausch, A. (2019). From the periphery to the center of global knowledge production? A bibliometric analysis of the evolution of a social science community from a small country: Austria. *Electron Journal, 34*, 12–19. https://doi.org/10.2139/ssrn.3475609

Tracy, J. D., & Chaudhuri, K. N. (1991). *Reflections on the organizing principle of premodern trade*. Cambridge University Press.

UNCTAD. (1999). *World Investment Report 1999*. https://doi.org/10.18356/8e79f24c-en.

UNCTAD. (2011). *World Investment Report 2011*. https://doi.org/10.18356/a3836fcc-en.

UNCTAD. (2012). *World Investment Report 2012*. https://doi.org/10.18356/5f5c0728-en.

UNCTAD. (2013). *World Investment Report 2013*. https://doi.org/10.18356/a3836fcc-en.

UNCTAD. (2021). *World Investment Report 2021*. https://doi.org/10.18356/a3836fcc-en.

Verbeke, A. (2003). The evolutionary view of the MNE and the future of internalization theory. *Journal of International Business Studies, 34*, 498–504. https://doi.org/10.1057/palgrave.jibs.8400062

Wallerstein, I. (2000). Globalization or the age of transition? *International Sociology, 15*, 249–265. https://doi.org/10.1177/0268580900015002007

Wallerstein, I. (2012). *World-Systems Analysis. 27*, 105–109. https://doi.org/10.1215/9780822399018-007.

Weede, E. (1985). *Entwicklungsländer in der Weltgesellschaft*. Springer.

Weiss, J. (1991). Export oriented industrialisation in developing countries. *Journal of Development Economics, 35*, 411–413. https://doi.org/10.1016/0304-3878(91)90060-9.

Part II
Globalization

Global Studies Meets World-Systems Theory

Manfred B. Steger

Abstract Honoring the significant academic lifework of Arno Tausch, this chapter addresses the strained relationship between the rising field of global studies and the more established framework of world-systems theory/dependency theory. Despite their common orientation toward worldwide social dynamics, world-system theorists have long been critical of globalization theory. However, maintaining a narrow focus on some differences between global studies and world-systems theory diverts attention from significant overlapping issues and common perspectives that connect these two paradigms. Arno Tausch's academic work is an instructive example of how various aspects of these two fields of global inquiry can be brought together in mutually beneficial ways. In the Austrian scholar's spirit of openness and reconciliation, this chapter briefly discusses the relevance of three shared features linking global studies and world-systems theory: the rejection of methodological nationalism; a strong inclination toward transdisciplinary; and a firm commitment to critical thinking.

1 Introduction

I first met Professor Arno Tausch in 1990 at the University of Hawai'i at Mānoa. Arno had arrived at the Hawaiian Islands to take up a semester-long appointment as Visiting Professor of International Relations at the Department of Political Science where I was completing my master's degree. As fellow Austrians and staunch cosmopolitans passionate about global affairs, we quickly became friends and spent significant time together on and off campus. What impressed me most about Arno then—and still does—was his impeccable scholarship; his encyclopedic mind that traverses several academic disciplines and areas of inquiry such as political economy, development studies, international relations, social science methodology,

M. B. Steger (✉)
Department of Sociology, University of Hawai'i at Mānoa, Honolulu, HI, USA
e-mail: manfred@hawaii.edu

G. Segell (ed.), *Development, Globalization, Global Values, and Security*,
https://doi.org/10.1007/978-3-031-24513-8_6

peace studies, and religious studies; his ability to connect his high-level quantitative research to normative concerns; and his highly ethical and open-minded personality that serves as the crucial link between his theoretical reflections and social engagement.

Luckily, Arno and I have maintained our personal friendship and collegial ties over the ensuing 32 years. Indeed, I am grateful to have been invited to contribute to this Festschrift which rightfully honors the academic lifework of one of the most significant political scientists Austria has produced in the last half century.

It is impossible to do justice to the full scope of Arno Tausch's wide-ranging scholarship in a chapter of merely four-thousand words mandated by the editor of this Festschrift. Hence, this essay confines itself to addressing a specific theme in Arno's weighty oeuvre that speaks to the status of globalization in the social sciences. Given my own expertise in globalization theory (Steger, 2008; Steger & James, 2019), this chapter focuses on the strained relationship between the rising field of global studies and the more established tradition of world-systems theory/ dependency theory. The latter, of course, constitutes the theoretical framework of much of Tausch's academic work. Yet, his ability to incorporate useful perspectives and analytic methods from both paradigms shows that global studies and world-systems theory need not be considered opposing approaches to the study of global processes.

2 Global Studies Versus World-Systems Theory?

Global studies emerged in the late 1990s as a transdisciplinary field of academic inquiry. It explores the many dimensions of globalization including economic, political, cultural, and ideological aspects. "What is there is also here and what is here is also there" is probably the most succinct and uncontroversial summary of globalization's central dynamics of interconnectivity, reconfiguration and compression of world-space and world-time, and enhanced mobility of people, goods, ideas, and institutions. Although globalization has been extensively studied in sociology, economics, anthropology, geography, history, political science, and other fields, it falls outside the established disciplinary framework that still bears the Eurocentric imprints of its academic origins.

Moreover, globalization is only of secondary concern in these traditional fields organized around different master concepts: "society" in sociology; "resources" and "scarcity" in economics; "culture" in anthropology; "space" in geography; "the past" in history; "power" and "governance" in political science, and so on. Conversely, global studies have placed "globalization"—a contested keyword without a firm disciplinary home—at the core of its intellectual enterprise. The rise of global studies in universities around the world represents, therefore, a clear sign of rising significance of understanding the expanding social interconnectivities and mobilities on a global scale (Appelbaum & Robinson, 2005; Steger & Wahlrab, 2017; Darian-Smith & McCarty, 2017).

World-systems theory is a generic label for neo-Marxist explanations of large-scale social change originating in the sixteenth-century Europe. Immanuel Wallerstein (1979, 2004), Giovanni Arrighi (1994), Christopher Chase-Dunn (1998), and other proponents of this approach adopted as their basic unit of analysis a single "world-system" rooted in structural components of the world economy. Driven by the extractive logic of capital accumulation, the capitalist world system combined with the nineteenth-century imperialist and colonialist Western practices to produce the unequal regions of the world system: the rich "core," the developing "semi-periphery," and the exploited "periphery." World-systems scholars present nation-states in the hegemonic global North not as the primary drivers of social change but as crucial servants of the overarching capitalist world economy.

Their principal function is to coordinate and reproduce the asymmetrical power relations that preserve the exploitative core-periphery hierarchy. During the 1960s and the 1970s, Latin American "dependency" thinkers like Osvaldo Sunkel (1969) and Fernando Enrique Cardoso (1979) developed a specific variant of the world-systems approach that was sharply critical of the linear and universalist approaches of economic development produced by dominant scholars in the global North. Perceptive European academics like Arno Tausch (2010, 2013) adopted this Latin American dependency perspective and proceeded to build their critical scholarship around pertinent themes of economic North-South inequities such as trade dependence, financial dependence, aid dependency, and so on.

Despite their common orientation toward worldwide social change, world-systems theorists have long been critical of global studies. Leading world-systems scholars like the late Immanuel Wallerstein (2004) and Christopher Chase-Dunn (1998) argue that globalization theory should be adapted to fit their own model, which suggests that the modern capitalist economy has been global since its inception five centuries ago. World-systems analysts reject the use of the term "globalization" as constituting a new master concept for a new field of inquiry that privileges the relatively recent compression of space and time over the last few decades.

Instead, they emphasize that the world-systems framework is more than adequate to analyze the globalizing tendencies that have been proceeding along the continuum of modernization for a long time. As Wallerstein (2004, x) observes, "The proponents of world-systems analysis. . .have been talking about 'globalization' since long before the word was invented—not, however, as something new but as something that has been basic to the modern world-system ever since it began in the sixteenth century." Prominent globalization theorists like Roland Robertson (1992) and Arjun Appadurai (1996) have pushed back, objecting to what they perceive as a narrow economistic framework and the corresponding neglect of cultural matters of the world-systems model.

As I argue in this chapter, however, an exclusive focus on real differences between global studies and world-systems theory diverts attention from significant overlapping issues and common perspectives that connect these two paradigms. For example, both global studies scholars and world-systems analysts emphasize the significance of developing new modes of analyzing the current transformations of world-space and world-time. In their view, serious discussions of transnational processes must pay attention to cross-regional transfers of resources, technology, and culture.

Extending their common historical sensitivities beyond the temporal spectrum of mainstream international relations (IR) scholars, both world-systems thinkers and globalization theorists have drawn on the insights of prominent world historians like Fernand Braudel (1982) and William McNeill (1964) to suggest that the origins of globalizing tendencies ought to be traced back to the political and cultural interactions that sustained the ancient empires of Persia, China, and Rome. Thus, the study of global dynamics should involve a single-frame analysis of total social systems over long periods of time.

This crucial sensitivity to world history is clear in Tausch's scholarship. Indeed, his academic work is an instructive example of how a variety of aspects of these two fields of global inquiry can be brought together in mutually beneficial ways. Thus, in the Austrian scholar's spirit of reconciliation, this chapter briefly discusses three of these shared features. First, both globalization theorists and world-systems analysts are opposed to forms of methodological nationalism that have long dominated the conventional approaches in both political science and IR.

Second, most representatives of global studies and world-systems theory embrace transdisciplinary approaches that challenge the defensive lines between economic, political, and sociocultural modes of analysis. Interested in capturing the "big picture," they emphasize that all forms of systemic knowledge necessarily involve the construction of overarching analyses and broad narratives that cut across existing disciplinary boundaries in their attempts to explain worldwide interconnections and social change. Third, both globalization theorists and world-systems scholars emphasize deep forms of critical thinking that address both analytical and normative issues.

3 Three Areas of Overlap

The first area of overlap between global studies and world-systems theory concerns their common rejection of methodological nationalism. This state-centric perspective still dominates much of the conventional IR subfield of political science, which constitutes Tausch's original disciplinary home. During the Cold War, both U.S. and European governments adapted methodological nationalism to fit what came to be known as "area studies," which was linked to Western-centric notions of modernization and economic development. Mainstream IR scholars treated the state as the main actor—and thus the central unit of analysis and the principal mover—of world politics in general.

Traditional IR scholarship focuses primarily on the self-interested actions of nation-states, especially regarding security issues and the balance of power. Methodological nationalism often combined with economic determinism at the expense of other crucial dimensions such as culture, ecology, and ideology. Attached to dominant behaviorist designs and rationalist modeling schemes, IR researchers tied to the Cold War areas studies paradigm sought to find generalizable analytical frameworks capable of explaining and predicting power dynamics in international affairs with the hope that such theories might contribute to the prevention of large-scale wars, especially a nuclear exchange between the West and the Soviet bloc.

By contrast, both global studies and world-systems theory developed an alternative intellectual agenda critical of the Cold War area studies paradigm. While recognizing the continued importance of the nation-state even in the post-Cold War context—and especially the rise of neoliberalism in the late twentieth and early twenty-first centuries—globalization scholars adopted flows, interdependencies, and transnational mobilities as its basic unit of analysis. This attention to the "global" became linked to what could be called "methodological globalism," a new framework which explores the many dimensions of worldwide interconnectivities.

The common rejection of methodological nationalism among global studies experts and world-systems theorists alike fosters a multi-referential understanding of a social whole that is no longer exclusively normalized within a national framework but is equally applicable to local and global settings such as global cities and regional networks of economic production. Ultimately, then, what makes a process or issue "global" has little to do with the common practice of dropping this trendy adjective in front of a growing number of nouns. Rather, the global comes to life in questions asked about the planetary dimensions of our existence that play themselves out in concrete settings across all spatial scales running from the local to global.

The second area of overlap between global studies and world-systems theory is reflected in their firm commitment to transdisciplinary. While the term of "interdisciplinarity" is quite familiar, the label "transdisciplinary" has emerged more recently in various academic settings. "Trans disciplinarity" signifies the systemic and holistic integration of diverse forms of knowledge by cutting across and through existing disciplinary boundaries and paradigms in ways that reach beyond each individual discipline. If interdisciplinarity can be characterized by the mixing of disciplinary perspectives involving little or moderate integration, then transdisciplinary should be thought of as a deep fusion of disciplinary knowledge that produces new understandings capable of transforming or restructuring existing disciplinary paradigms. But the transdisciplinary imperative to challenge, go beyond, transgress, and unify separate orientations does not ignore the importance of attracting scholars with specific disciplinary backgrounds. Rather, transdisciplinary teams typically use a shared conceptual framework and draw together discipline-specific theories, concepts, and approaches to address a common problem (Choi & Pak, 2006, 355).

Challenging the Eurocentric disciplinary framework shaped by the national imaginary of the nineteenth and twentieth centuries, both global studies scholars and world-systems theorists embrace multiple forms of transdisciplinary. These involve, for example, the formation of issue-driven teams of both academic investigators focused on real-world problems like violence, inequality, and environmental degradation and dedicated to the integration of knowledge involving several disciplines such as sociology, urban studies, political science, geography, anthropology, history, biology, engineering, ecology, and economics. Both camps agree that the transdisciplinary study of transnational flows and interdependencies requires paying attention to close encounters with multiple forms of complexity and differentiation.

After all, the prevailing compartmentalization of scientific and professional knowledge reflected in the modern disciplinary order fails to capture the increasingly diverse and multilayered social contexts of the twenty-first century. By recognizing the importance of increasing complexity for their systematic inquiries, global studies experts and world-systems theorists alike consciously embrace transdisciplinary in their efforts to understand the shifting dynamics of interconnectedness. Thus, their exploration of complex forms of interdependence not only combats knowledge fragmentation and scientific reductionism, but also facilitates an understanding of the "big picture," which is indispensable for stimulating the political commitment needed to tackle the pressing global problems of our time such as large-scale human migration, proliferating digital technologies, and novel pandemics such as Ebola and COVID-19.

Since global complexity appears in many forms and pervades different social arenas, it makes sense for global studies scholars and world-systems analysts to approach it from different angles and through multiple levels of analysis. The goal is to illuminate the intensifying complexities involved in the interconnectivity of microscopic, "local" phenomena and macroscopic dynamics involved in the creation of globality. Grasping these "glocalized" forms of interdependence requires researchers to experiment with innovative methodological approaches that reject the local/global binary (Robertson, 1992; Roudometof, 2016). This calls for the conscious application of a non-reductionist account of interactions across multiple, often overlapping, spatial scales and dimensions at which social action takes place in our globalizing world.

The third area of overlap between global studies and world-systems theory concerns their strong commitment to employing and encouraging various forms of critical thinking. Going beyond the purely cognitive understanding of "critical thinking" as "balanced" and "detached" reasoning propagated by leading Anglo-American educators during the second half of the twentieth century, this stance reflects both fields' receptivity to the activity of social criticism that problematizes unequal power relations and engages in ongoing social struggles to bring about a more just global society. Advancing various critical perspectives, global studies and world-systems scholars from around the world draw on different currents and methods of "critical theory." Note that this term is no longer confined to the intellectual tradition of the Frankfurt School but manifests today as an umbrella concept for multiple modes of thought committed to the reduction of exploitation, commodification, violence, and alienation. Such a critical approach to complex transnational processes calls for the steadfast contestation of uncritical mainstream stories spun by corporate media, the decolonization of the Eurocentric imagination, and an understanding of the global as a multipolar dynamic reflecting the concerns of the marginalized global South even more than those of the privileged North.

Taking sides with the interest of social justice, many critical thinkers affiliated with both global studies and world-systems theory exercise what globalization scholar William Robinson (Appelbaum & Robinson, 2005, 14) calls a "preferential option for the subordinate majority of global society." The globalization scholar James Mittelman (2005, 19–25) offers an important elaboration of this common critical framework connecting global studies and world-systems theory. His coherent complex of critical knowledge consists of the following five interacting components.

The first is "reflexivity," which connotes an awareness of the relationship between knowledge and specific material and political conditions. Mittelman insists that to be reflexive means to probe the historical context and power interests embedded in various globalization perspectives. This critical emphasis on often-conflicting historical perspectives also relates to Mittelman's second component—historicism—which incorporates the time dimension in the critical study of transnational processes. Rigorous historical thinking has the potential to correct abstract views portraying specific forms of globalization—such as the global integration of markets—as the inevitable outcome of a benign capitalist process. Mittelman refers to the third component as "decentering," which involves the production of myriad perspectives and forms of knowledge of globalization from both its epicenters and the margins.

Such a "decentered lens" provides the necessary peripheral vision that enables critics to see global dynamics from multiple angles and spatial locations—especially from the point of view of the marginalized regions of the global South. Mittelman characterizes transdisciplinary, the fourth element of his critical framework, as the ability to forge crossovers between the social sciences and complementary branches of knowledge. Key to a critical understanding of globalization, crossovers encourage scholars to break through disciplinary barriers in their holistic pursuit of real-world problems. Strategic transformations, the final component of Mittelman's framework, involve a willingness to present ongoing challenges to the hegemonic power interests of the incipient global society.

For Mittelman, the organic intellectuals behind these strategic maneuvers develop emancipatory visions that have practical purposes for the formation of more democratic forms of globalization. As he notes, "The goal is to inculcate a new moral order in lieu of the dominant ethos—currently an ethos of efficiency, competition, individualism, and consumption inscribed in neoliberalism" (Mittelman, 2005, 24–5). Indeed, Arno Tausch's persistent scholarly efforts over several decades to expose the shortcomings of neoliberal market globalization and its deeply adverse effects on the global South exude the same critical sprit reflected in Mittelman's suggested framework (Tausch & Heshmati, 2011, 2012; Tausch, 2016).

4 Concluding Remarks

I would be remiss not to acknowledge the existence of some clear differences between the world-systems and global studies outlook. While both approaches dismiss the utility of making the nation-state as the basic unit of analysis, world-systems analysts focus their research efforts on "world-systems" rather than globally circulating flows and interconnectivities. In fact, Wallerstein insists that his use of the term "world-system" should not be equated with the idea of a "global system." As he puts it, "Note the hyphen in 'world-systems' and its two subcategories, 'world-economies' and 'world-empires'. Putting in the hyphen was intended to underline that we are talking not about systems, economies, empires of the (whole) world, but about systems, economies, empires that are a world (but quite possibly, and indeed usually), not encompassing the entire globe" (Wallerstein, 2004, 16–17).

Secondly, Wallerstein and other prominent world-systems theorists leave little doubt that they consider world-integration to be a process driven largely by economic forces whose essence can be captured by analytical models focusing on capitalist dynamics. Accordingly, they assign to culture and ideology merely a subordinate role as "idea systems" dependent on the "real" movements of the capitalist world economy (Wallerstein, 1990, 38). By contrast, only very few global studies researchers analyzing the economic dimensions of globalization subscribe to similar forms of reductionism.

To be fair, however, it is necessary to point out that a growing number of studies produced by such prominent world-systems scholars as Barry K. Gills (2008), Leslie Sklair (2002), and Andre Gunder Frank (1998) acknowledge the interaction between dominant-class interests and cultural transnational practices. In so doing, they have begun to raise important normative questions, suggesting that the elements of the "ideological superstructure"—politics, ideas, values, and beliefs—may, at times, neutralize or supersede economic forces. Sklair (2002), for example, examines transnational class dynamics that highlight the importance of what he calls "the culture-ideology of global consumerism."

Other world-systems thinkers like Ash Amin (1997) concede that the pace of globalization has significantly quickened in the last few decades of the twentieth century. This unprecedented compression of space and time in recent years suggests that much of the world-systems criticism of globalization as a new phenomenon has been based on quantitative analyses of trade and output that neglect these recent qualitative shifts in social and political relations. These spectacular changes in the globalizing process, Amin (1997) argues, may indeed warrant the construction of a new research paradigm known by the label "global studies" and dedicated to the analysis of new configurations in the capitalist world-system such as complex networks of transnational corporations, financial flows, global supply chains, and cultural homogenization.

And it is precisely at this crucial point of today's increasing academic recognition that various analytical tools and perspectives from global studies and world-systems theory can be employed in a complementary fashion where Arno Tausch's valuable contribution fits in. His analysis of globalization is anchored in the skillful employment of all three overlapping issues discussed in this chapter. First, his refusal to adopt the nation-state as his basic unit of analysis is reflected in his choice of article titles which frequently employ the terms "cross-national" or "transnational" (Tausch, 2011; Tausch & Heshmati, 2012). Second, his commitment to transdisciplinarity is impressively demonstrated in the placement of his work in journals and other publishing outlets that serve multiple disciplines and fields of inquiry. Third, his critical spirit becomes visible in his sharp analyses of neoliberal globalization as leading to increasing levels of inequality and fostering North-South disparities in wealth and wellbeing (Tausch, 2010, 2013, 2016; Tausch & Heshmati, 2012).

Hence, his call for the implementation of more ecological sustainable and socially responsible global development strategies (Tausch & Heshmati, 2012, xxxix). Indeed, Arno Tausch's work constitutes not merely a successful synthesis of world-systems and global studies approaches, but also displays its author's deep emancipatory interest in an ethic of North-South cooperation that is indispensable to making our globalizing, but increasingly unsettled, world of the twenty-first century a better place for all sentient beings.

References

Amin, A. (1997). Placing globalization. *Theory, Culture & Society, 14*(2), 123–138.

Appadurai, A. (1996). *Modernity at large: Cultural dimensions of globalization*. University of Minnesota Press.

Appelbaum, R. P., & Robinson, W. I. (2005). *Critical globalization studies*. Routledge.

Arrighi, G. (1994). *The long twentieth century*. Verso.

Braudel, F. (1982). *On history*. University of Chicago Press.

Cardoso, F. E., & Enzo, F. (1979). *Dependency and development in Latin America*. University of California Press.

Chase-Dunn, C. (1998). *Global formation: Structures of the world economy*. Rowman & Littlefield.

Choi, B., & Pak, A. (2006). Multidisciplinarity, interdisciplinarity, and transdisciplinarity, education and policy: Definitions, objectives, and evidence of effectiveness. *Clinical Investment, 29*(6), 351–364.

Darian-Smith, E., & McCarty, P. (2017). *The global turn: Theories, research designs and methods for global studies*. University of California Press.

Frank, A. G. (1998). *ReORIENT: Global economy in the Asian age*. University of California Press.

Gills, B. K. (2008). *Globalization and the global politics of justice*. Routledge.

McNeill, W. C. (1964). *The rise of the West: A history of the human community*. University of Chicago Press.

Mittelman, J. H. (2005). What is a critical globalization studies? *Critical Globalization Studies, 19*(1), 32–45.

Robertson, R. (1992). *Globalization: Social theory and global culture*. Sage.

Robinson, W. I. (2005). What is a critical globalization studies? Intellectual labor and global society, *Critical globalization studies*. 19(2), 11–18.

Roudometof, V. (2016). *Glocalization: A critical introduction*. Routledge.

Sklair, L. (2002). *Globalization: Capitalism and its alternatives*. Oxford University Press.

Steger, M. B. (2008). *The rise of the global imaginary: Political ideologies from the French Revolution to the Global War on Terror*. Oxford University Press.

Steger, B., & James, P. (2019). *Globalization matters: Engaging the global in unsettled times*. Cambridge University Press.

Steger, M. B., & Wahlrab, A. (2017). *What is global studies? Theory and practice*. Routledge.

Sunkel, O. (1969). National development policy and external dependence in Latin America. *The Journal of Development Studies, 6*(1), 23–48.

Tausch, A. (2010). Globalization and development: The relevance of classical "dependency" theory for the world today. *International Social Science Journal, 61*(4), 467–488.

Tausch, A. (2011). Is globalization a driver of economic growth and development? A cross-national, quantitative perspective on current, pro-globalist strategies of the European Union. *Society and Economy, 33*(3), 505–524.

Tausch, A. (2013). Reconsidering migration, globalization and social conditions in the world system. *Bogaziçi Journal of Social, Economic and Administrative Studies, 27*(2), 63–82.

Tausch, A. (2016). Is globalization really good for public health? *The International Journal of Health Planning and Management, 31*(2), 511–536.

Tausch, A., & Heshmati, A. (2011). Re-Orient? Understanding contemporary shifts in the global political economy. *Journal of Globalization Studies, 2*(2), 89–128.

Tausch, A., & Heshmati, A. (2012). *Globalization, the human condition, and sustainable development in the twenty-first century: Cross-national perspectives and European implications*. Anthem Press.

Wallerstein, I. (1979). *The capitalist world economy*. Cambridge University Press.

Wallerstein, I. (1990). Culture as the ideological battleground of the modern world system. *Global Culture, 9*(3), 31–56.

Wallerstein, I. (2004). *World-systems analysis: An introduction*. Duke University Press.

World-Scale Problems and the Policy Response of Multilateral Institutions

Fulvio Attinà

Abstract This delves into world-scale problems addressed by Arno Tausch and the policy response of multilateral institutions. The first section reviews the existing knowledge about multilateral policymaking to address world-scale problems. The second section discusses the application of multilateral policymaking to three current world-scale problems, namely climate warming, pandemics, and mass migration.

1 Introduction

Collective, world-scale problems gripping the world today should be addressed with the consistent response of all or almost all states, including large and strong states and small and weak ones. But sovereignty and inequality, two fundamental features of the world political structure, make it difficult to achieve this condition. Collective action problem thinking provides an understanding of why governments most of the time prefer to go their separate ways to respond to collective, world-scale problems instead of coordinating their response. Governments pursue individual, bilateral, and multilateral responses to address problems regardless of scale. However, the establishment of international organizations since the late nineteenth and early twentieth century to respond to transnational problems under the impact of the first and second industrial revolution demonstrates that the states are not strangers to recognize the advantages of tool standardization, resource pooling, and even policy convergence to address collective problems.

Furthermore, on the wake of the Second World War, they made the first significant decision to using policy convergence to respond to collective problems. The post-WW II conferences of the victorious states created the first policymaking institutions that formed world "public" policies to address four collective problems which were considered crucial to restoring normalcy in world politics. Reducing

F. Attinà (✉)
University of Catania, Catania, Italy
e-mail: attinaf@unict.it

G. Segell (ed.), *Development, Globalization, Global Values, and Security*,
https://doi.org/10.1007/978-3-031-24513-8_7

international trade tariffs, giving stability to the world finance and national currencies, saving all states from lethal armed aggression, and replacing states to colonies were the problems recognized as crucial to address by forming specific world policies. Multilateralism was the name given to the policymaking method of the institutions that the coalition of the Western countries created for such a goal under the leadership of the USA. These countries won the World War in alliance with the Soviet Union but against the will of the Soviet Union they were able to form those world policies that also created the new world political order.

Multilateral policies thus became the way around the problem of collective action in world politics, although it is not always possible to profit from this method of policymaking. The post-World War II conditions were favorable for the purpose. They are not as much good today. In today's world, after the delegitimization of the four initial world policies, multilateralism faces the mistrust of the policymakers that are centered on state sovereignty even though new, severe worldwide problems are pressing to be addressed properly. The reason is that world order has entered the transition mode.

The present paper aims at understanding policy response to world-scale problems in such circumstances that are deeply different from the conditions at the time multilateralism entered world politics. The first section reviews the existing knowledge about multilateral policymaking to address world-scale problems. The second section discusses the application of multilateral policymaking to three current world-scale problems, namely climate warming, pandemics, and mass migration. Such problems are very different in nature, and, above all, they are at different stages of recognition as problems to be processed by multilateral institutions. The Conclusions section resumes the chapter argument.

2 Multilateral Policymaking

Giving an international institution the power to form the world policy toward a problem in a sovereignty sensitive area is the good choice if two conditions are met: first, member states recognize the political authority of the institution and, second, policy is accepted as legitimate to address a problem whose origins and effects are recognized as being of a worldwide scale. The first condition is achieved by the signature and ratification of the statutes of the institution by the designed national institutions. The second condition is more complex to ascertain. However, for a reason that will be made clear later in this chapter, it can be said that, today, such legitimacy relies on the principle of national ownership.

Such principle includes both ensuring equal rights of policymaking to all the states and implementing the world policy through national policies that protect the culture and interest of each state. These are not insurmountable conditions. The key question is which problems the states recognize as collective problems to be addressed through world policymaking institutions. Environmental pollution was the main issue raised on the world stage in the early Seventies when the United Nations took on the responsibility of organizing conferences to push policymaking toward responding to world-scale problems. Generally, the UN conferences that address collective problems end with the signature of political declarations, sometimes with international law treaties that put obligations upon the states. Normally, such conferences refrain from forming a world policy, that is, a consistent set of principles, rules, and programs to address collective problems through organizational, human, and financial resources they states are called to comply with.

In general, a problem creates a condition of harm to be addressed to regain control over one's environment. If the problem is not solvable, it is addressed through mitigation and adaptation measures. Mitigation measures reduce the level and number of adverse effects of the problem. Adaptation measures increase the ability of actors to resist the negative effects of the problem. Collective problems harm the whole society and every member. The effective response to such problems is at hand if all members give, on their own, compatible responses or agree to form the collective response that has the form of a framework policy and subsequently establish their own policy consistent with the collective one. Otherwise, as predicted by the studies on the problem of collective action, the heterogeneous responses of social actors do not achieve efficient and effective results, but collide and harm each other. On such premise, in world affairs, the multilateral policymaking can be the way out of the collective action problem.

Multilateralism is the name of many forms of international cooperation. Regular intergovernmental consultation such as the G7 and G20, actions of small groups and large coalitions, the negotiation of international treaties, and the actions of international organizations are examples of what the media and experts call multilateralism. All such cooperation forms are significant and different from one another and as well from what was named multilateralism at the time the post-war conferences. In 1945, the San Francisco Conference created the United Nations Organization and charged it with various tasks, including to respond to the problem of saving all states from fatal aggression and the problem of ending colonialism.

The Bretton Woods Conference established the International Monetary Fund and the World Bank which developed the policy toward the problems of the currency exchange rates and the stability of world finance. Last, a conference approved the General Agreement on Trade Tariffs (GATT) that formed the world policy toward the gradual reduction of trade tariffs. Over the next seventy years, these institutions and policies underwent major changes and met with the opposition of the dissatisfied states.

Delegitimization and the consequent diminishing effectiveness of such policies lowered confidence in multilateral decision-making and influenced the world political order as well. In last years, the delegitimization of the multilateral institutions and policies has grown (Tausch, 2010; Tausch & Heshmati, 2011). Both skeptics and proponents of multilateralism agree on its poor state of health. Contrary to such view and based on the lessons of history, the present section defends multilateralism as the fit-for-purpose policymaking mechanism useful to build world policies to respond to the far-reaching problems that today affect all states.

In the following section, the model of the post-WW II multilateral policymaking is briefly reminded and compared with the model of the world policymaking toward the climate warming problem, the most significant case of the today's irreplaceability of multilateral policymaking to respond to world-scale, collective problems.

After studying the multilateral policymaking of the post-WW II conferences, John Ruggie, the founding author of multilateralism research (1993), pointed out the three conditions of multilateral policymaking. They are the initial consensual selection of the policy principles, the equal decision-making rights of the states during the process, and non-discrimination in policy implementation. Though such conditions are essential to policy legitimacy, the post-WW II conferences that created multilateral policymaking institutions did not used them in full and in all cases but adopted two models of policymaking, the top-down and bottom-up model.

The former allocates the primary authority of policymaking to the institution, gives preferential decision-making rights to some countries, and let the policy implementation to the individual state though giving conditional support to the state in need of assistance. It is the model of the world finance policy and the state security policy that was imposed by the Western countries at the Bretton conferences. The bottom-up model, on the other hand, assigns primary authority to the individual members of the institution and does not attribute a role to the institution in policy implementation. It is the model of the world trade policy that GATT has formed, and the WTO has continued.

In 2015, the world climate policy has been agreed with the signature of the Paris Agreement that came after a long process that started in 1992 when the United Nations Framework Convention on Climate Change, UNFCCC, was approved. The Paris Agreement created the third model of multilateral policymaking, here named circular or loop model. It complies with the three Ruggie's conditions as well with the principle of the state ownership of the policymaking and implementation under the monitoring and supervision of the policy institution. Such model puts the UNFCCC in the position of guiding the reform of the policy in association with the member governments. Last, non-state and substate stakeholders participate in the policy process, apart from voting the official documents.

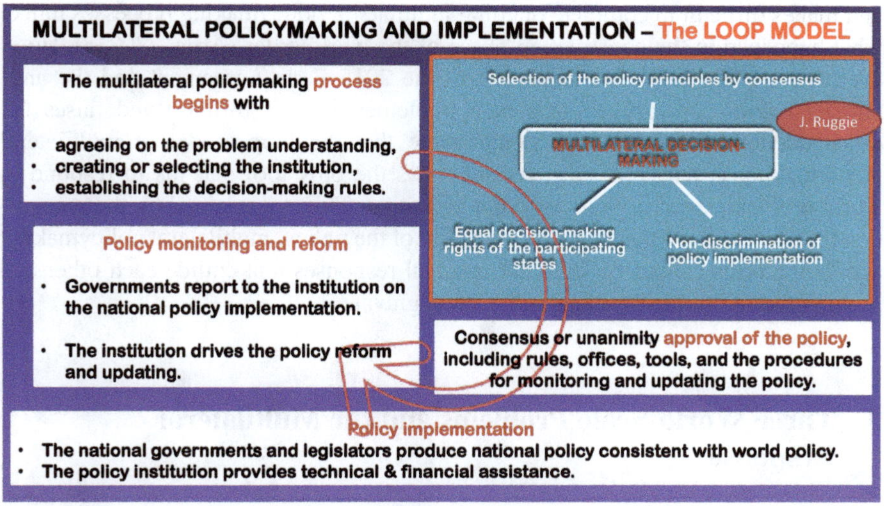

Multilateral policymaking has become a double level and circular process that, based on the founding international law treaty, associates all the states with equal decision-making rights in the formation of the world policy, which has the form of a framework policy. The agreed set of principles, rules, and programs to address the collective problem that, thanks to organizational, human, and financial resources, is implemented by the individual countries through national legislation and policies. All governments report about the implementation to the policy institution, which will support the implementation capacities of the states in need of assistance and can drive the reform of the policy.

Many researchers have assessed such policymaking process focusing on crucial aspects such as the principle of differentiated implementation and the distribution of the costs and resources of the policy, the definition of the policy paradigm, that is the shared understanding of the nature and causes of the problem that predetermines the policy goals and instruments (Coleman, 2021), the problem of the uncertainty about the effects of the policy on the parties (Carr & Lesniewska, 2020), and the instruments for monitoring and checking the compliance with the policy by the individual state (Ivanova & Escobar-Pemberthy, 2021).

In closing the present section, I argue that multilateral institutions and policies also are instruments of world political order. Since they have an impact on the internal affairs and foreign relations of the states that commit to produce domestic policies consistent with the world policy, world order is strengthened by the homogeneous response of the states to the problems of common interest. Conversely, when the number of states dissatisfied with existing multilateral policies grows and the world policies either are not transposed into domestic policies or cease to be perceived by the states as binding, the delegitimization of the institutions and policies follows and as a result the world order declines.

The delegitimization of the institutions and founding policies of the American world order has created the current world disorder. Also, it affects multilateralism and makes difficult to complete ongoing multilateral policymaking processes unlike what happened in the construction phase of the world order (Attinà, 2021a). However, the signing and implementation of the 2015 Paris Agreement and the anti-climate warming policy, though such implementation is difficult and raises the dissatisfaction of some states, demonstrate that the loop model of multilateral policymaking is received by governments as the only tool that fits to respond to world problems even in times of order transition. This is explained by the recognition that, by ensuring the national ownership of the policy, multilateral policymaking avoids the costs caused by different national responses that collide each other and prolong the life of the problem and even aggravate it.

3 Three World-Scale Problems and the Multilateral Response

Although the effectiveness and legitimacy of post-war multilateral policies are weakened, new problems push to resort to multilateral policies. Disasters caused by climate warming, pandemics, and forced migration are example of such problems. They are already the responsibility of multilateral institutions, in particular the UNFCCC for climate change, the WHO for health, and the UNHCR, IMO, and other international organizations for forced migration. These problems and the action of the relevant institutions put before the national governments the dilemma of prioritizing national response or recognizing that the national response must be consistent with that which is formed within multilateral institutions with the equal participation of all the member states.

The present section draws attention to the state of the multilateral response to such problems. The interest of political scientists in such problems is high and increasing. The UNFCCC and the world climate policy are much researched. The WHO is receiving growing attention because of the COVID 19 pandemics. Migration has been the object of political research for the past three decades, but the current wave of mass migration is not perceived as a problem to be dealt through a multilateral policy. The UN Compact on Migration process is understudied by political scientists.

The three problems are briefly described below, and knowledge about the current state of the world policy response toward such problems is presented. The aim is to emphasize the need for research to build comparative knowledge on multilateral policymaking toward world-scale problems.

4 Climate Warming

In the 1970s, environmental scientists warned about the negative effects of the overconsumption of fossil fuels. The CO2 emissions of fossil fuels add an excessive amount of gas to the earth's natural gases and change the beneficial greenhouse gas (GHG) effect that protects against the heat emitted by the sun on earth. As a result, the earth temperature grows and the level of the seas, the intensity of rainfall, and the life of humans, animals, and plants are badly affected. Since then, the states have posed themselves the problem of climate warming. They concluded that climate change is the perfect collective problem of world reach because it affects everyone and must be addressed by all states with shared principles, objectives, and policy standards (Tausch & Heshmati, 2018). In 2015, the national governments achieved the collective response through the consensual approval of the treaty known as the Paris Agreement.

The Paris Agreement is the result of the long process of multilateral policymaking that started in 1992 with the signature of the United Nations Framework Convention on Climate Change and the following birth of the world climate policy institution named UNFCCC, the acronym of the signed document. The Agreement, that has been ratified by all the states apart from Iran, Iraq, and Libya, engages the governments to form national policies consistent with the objectives, standards, and time targets of the world climate policy.

Every two years, the UNFCCC general meetings, known as COPs (Conferences of the Parties), give to states and non-state stakeholders the right and voice to bring their values and interests into the policy. The policy goal is reducing GHG emissions by replacing fossil fuel energy with renewable energy sources such as wind, solar, and other non-polluting sources and by means of carbon offsetting measures. The states committed to hold climate warming to well below 2 °C, though experts call for committing to the 1.5 °C target, and to decarbonize the world by 2050 to avoid the risk of reaching 3 °C or higher warming by 2100.

The policy implementation relies on the national ownership of the policy and the UNFCCC's monitoring and assistance responsibility. Every five years governments agree to submit a report to the UNFCCC Secretariat. The report contains the Nationally determined contributions (NDCs) that make public what each state does to converge to the world policy goals. States recognize that responsibility for causing climate warming is common but differentiated. Consequently, they share the principle of common and differentiated convergence because the financial and technological resources necessary to implement policy are owned by states to varying degrees. UNFCCC bodies are tasked with organizing financial and technical assistance to strengthen the capacity of the states that need external support to implement the policy. Non-state stakeholders support the process with knowledge, advice, independent monitoring, and other resources. For this reason, experts use to talk of world climate governance, but at the core of the world response is the multilateral policymaking that is owned by the state governments.

The world climate policy is unrolling as a work in progress whose model is new and clearly shaped (Abdenur, 2021; Attinà, 2021b). In fact, the response to the climate warming problem is reached through the circular policy model that has been described earlier in the present chapter. The state governments give to the world institution the responsibility and power of forming the policy through the multilateral policymaking method. Thereafter, each government implements the policy by passing consistent national legislation and policy. Finally, the world institution monitors implementation by states, assists States in need of help, and guides policy updating if necessary.

China, the USA, the European Union, India, Russia, and Japan are the countries on top the rank of GHG emissions (Crippa, 2020). In 2019, their GHGs share has been 66.9% of the total emissions, varying from Japan's 3.0% share to China's 30.3% share. A group of 10 countries complete the ranking of the emitters with more than 1% of the total fossil CO2 emissions in 2019. Together, the 16 countries account for 81.2% of world emissions. Achieving the world climate policy goals is the responsibility of the governments of these and all the countries. They have agreed on mandating the formation of the policy response to the world policy institution by applying the multilateral policymaking method. However, the results of the policy depend on the domestic politics and constraints of the states that must implement the policy by converging on the world policy.

Two internal conditions most affect the national propensity to policy convergence, the per capita GDP and the production and export of fossil fuels. The former signals the importance of economic development that should not be harmed by the decarbonization policy. Such a condition is taken into consideration through the principle of differentiated convergence. The latter condition is harder to cope with. One expects that the countries with the largest reduction of CO2 are not the producers and exporters of fossil fuels. The reality is different. Some of the countries with the largest reduction are the largest fossil fuel producers, namely China, the USA, Russia, Canada, and Australia. But Saudi Arabia and Iran, the two countries with the highest ratio of CO2 per GDP, are big fossil fuel producers. The fossil-fuel-rich countries will suffer big capital losses as their reserves are left unexploited for mitigating the climate change.

They should engage themselves in shifting to produce renewable energy resources. Otherwise, they will slow down the implementation of the world climate policy to gain time to adapt their economy to energy transition. However, the economic and technical conditions are at hand (Hafner & Tagliapietra, 2020). Substituting renewable energy sources to carbon energy sources is feasible and affordable since these sources are abundant. The problem is political and certainly a hard world political problem at the time of order transition. The actual President of the USA, the country rich in fossil fuels and the most politically important country in the world, has brought his country back in the Paris Agreement policy. The American role in the climate policy should impact also on the world order transition by pressing on China whose green policy, acknowledged by all, is not as determined as it should be. China's leaders have said that reducing GHG emissions within 2030 is not fully compatible with the country's present development strategy. The war in

Ukraine is the blatant demonstration of the relation between the world climate policy and the world political order transition. Such a relation is of high importance to multilateralism and will have an impact on the policymaking process to respond to other world-scale problems.

5 Pandemics

Different levels of economic development and different mentalities and habits condition the health care policy of the states. The elimination of differences in the right to health is a complex problem that should be addressed to break the poverty-disease-poverty circle. When many people of a country are afflicted with diseases, the economic and social development of the country is conditioned and the problems that are caused by underdevelopment spread to the outside world. Ultimately, the collective response to health problems in every corner of the world is in the self-interest of every state as much as an ethical and humanitarian duty. This applies especially to infectious diseases. The more limited is the public health budget, the smaller the state's provision of health care to citizens and the capacity to respond to infectious diseases. Since states are powerless to stop the spread of infectious diseases across national borders also because of the easy and intense movement of people in the contemporary world, epidemics turns easily into pandemics, i.e. into a world-scale problem (Tausch, 2012, 2015).

As is the case with other world-scale problems, the effectiveness of the world response to the problem of the infectious diseases depends on three requirements: attributing to an international organization, which today is the World Health Organization (WHO), the responsibility of acting as the multilateral policymaking institution, forming national health policies consistent with the world policy toward communicable diseases, and undergoing institutional monitoring and scrutiny to ensure the true implementation of the policy.

WHO provides technical assistance and disseminates knowledge to improve national health programs. On a higher level, WHO produces rules on health questions through decisions of its organs. In 2005, the WHO Assembly approved the second edition of the International Health Regulations (IHRs), entered into force in 2007. In consideration of the growth in international travel and trade and the emergence or re-emergence of international disease threats and other public health risks, the IHRs contain norms about the notification of epidemiological information and other procedures and sanitary measures the national health authorities must observe. Formally binding, but non-enforceable, the IHRs are the object of poor compliance by states who "privilege sovereignty over granting the WHO the requisite authority to fulfill its mandate" (Kreienkamp & Pegram, 2021).

COVID-19 has made clear that the worldwide control of infectious diseases is a world-scale problem and, as the experts say, a world public good. Brown and Susskind remark that the national governments ensure the optimal level of a public good to their citizens, but "there are no provisions in international law to impose obligations on other sovereign countries to ensure the optimal overall level of a GPG (global public good) is secured—without their consent" (2020, S65). In truth, an international law treaty imposing obligations on national governments is essential but not sufficient condition. The international law treaty should also establish the world policy toward the infectious diseases. It should be negotiated and approved through a multilateral process that can be assigned to the WHO and empower the WHO to form the policy, monitor the implementation by the states, and assist the states that lack the appropriate implementation capacity.

Brown and Susskind also remark that controlling an infectious disease worldwide involves many tasks that "may be 'good' activities" (2020, S66) for the whole of the states.

> "To begin with, there are 'best-shot' public goods, where the amount of the public good supplied depends on the largest contribution made by an agent. Vaccine discovery, for instance, is a best-shot GPG: whether a vaccine is found strongly depends on the particular countries that contribute the most to its discovery. Then there are 'weakest-link' public goods, where the amount of the public good supplied depends on the smallest contribution made by an agent. Preventing the virus spreading, for example, is a weakest-link GPG: so long as there is a single country that fails to control the disease within its own borders, it is possible that the virus could spread beyond its borders and around the world. Then there are 'threshold' public goods, where some critical level of contribution must be made by each agent before the public good is produced .... Disease surveillance, for instance, is a threshold GPG: only once a critical number of countries or regions have tracking systems in place is it possible for the global community to understand the epidemiology of the virus" (Brown & Susskind, 2020, S66–67).

The world policy toward the control of infectious diseases should be devised to respond to such issues as ensuring that the efforts to provide control are well coordinated among the states and the countries that lack the capacities to contribute to controlling the infection spread are supported by the others. The response to the COVID-19 pandemic has made clear how difficult it is to adequately face such issues. Independent research programs and vaccine nationalism took place. Support to less developed countries unable to control the spread of the disease on their own has been insufficient, to say the least. Transparency failures, from the China's delayed release of fundamental data to the poor information on the national task forces programs, have been numerous. On top, the WHO has been the object of delegitimizing criticism that has undermined her capacity of action as the only existing institution responsible for coordinating the response to the virus. The WHO request of funds continues to be largely unaccounted.

The Trump administration took the lead of the WHO de-legitimation by terminating the country membership after suspending funding to the organization, and by opting out of the Covax initiative that aims to ensure equitable access to COVID-19 vaccines. Though the Trump administration was not alone in questioning the WHO, other countries including the United Kingdom, France, Germany, Japan, and China increased financial contributions and backed Covax. The Gates Foundation, the second largest contributor to the WHO's budget after the USA, also made extra funding available.

President Biden reversed the exit decision, recommitted the USA to the agency, and provided support for WHO-led activities such as Covax. The WHO has been defended by many state and non-state actors that have taken for granted her instrumental, health-scientific legitimacy (Yang, 2021). It served as a central hub for information, expertise, and technical assistance, in a field in which organized expertise and science-based policy are of critical importance. Reforming the WHO should be good as much as the objective is boosting her authority as a technical-scientific institution and shared policymaking responsibility with the member countries.

6 Mass Migration

The world-systemic sources of migration attract much less attention than the conditions of the migration origin countries. Few believe that the drivers, that is the conditions causing, incentivizing, and facilitating mass migration in today's world, are the result of interrelated international and national factors. The structure of the world economy, which is sustained by international law treaties and the world policies of finance, production, and trade, the action of international organizations that directly and indirectly deal with migration, the human rights ideology, the intermediary action of the NGOs, the impact of climate change technology, and the synergy of all these and other aspects of the contemporary world are scored as much less important than domestic conditions such as ethnic violence, civil conflict, political repression, bad governance, corruption, and, above all, the backward stage of the national economy of the migration origin countries.

Migration has been always considered a normal social phenomenon. Human beings settle in the place of birth forever or migrate. The migrant either goes back to own country or delocalizes him/herself for good. But, on occasion, migration changes into an unusual phenomenon, the migration wave. A huge number of persons crowd the routes out of a certain area. Age of migration is another way of saying that for a long period stable conditions inexorably cause large migration flows. When such a case occurs, the policymakers of the inflow countries should allocate resources to receive numerous migrants and keep normal the country life. If this does not happen, the government's response falls dependent on the citizens' perception of the migrants that normally is perceiving migrants as a threat to the inflow country's normal life.

For almost a decade now, the EU countries, the USA, and Australia are the case in point. Africa and large parts of Latin America and of Central-Southeastern Asia are the areas of origin of the wave. Briefly, migration is a world-range phenomenon also taking into account the inward and outward migration of the Chinese people be taken into consideration. In fact, in the early 1990s, migration experts warned that the world had entered the age of migration. Demography experts called attention to the flywheel effect of the mix of high birth rate, low income, and no job opportunities in

Africa. Not to mention the alarm raised by the economists for the unstoppable unemployment rate in the countries of the periphery of the world.

On such assumption, the baseline argument of the present section is that the current migration wave is caused by interrelated conditions and trends taking place at the world and local level (Attinà, 2016; Tausch, 2014). In the common view, instead, the causes of migration are rooted in the country of origin. The "causes" of the destination, the so-called pull factors, are the opportunity differentials between the migrant's country and the preferred country of destination. The phenomenon is the result of the combination of drivers that involve all the countries of the world. Accordingly, the current migration wave is a world-scale problem and is perpetuated by addressing it through bilateral agreements between sending and receiving countries, by charging specialized international organizations with the task of assisting the migrants, and by forming international law rules that put upon state governments above all humanitarian and human rights obligations toward migrants.

The less than sub-optimal results of such response do raise problems to individual states that defend their individual sovereignty, and to the whole world. In such circumstances, the "self-irreplaceability" of the world drivers and the unmanageability of the local drivers by the rulers and citizens of the countries of origin give to the current migration the nature of a problem to address as a world-scale problem, i.e. through a world policy formed and implemented through the circular multilateral model. The decade-long experience of the American, European, and Australian governments that have activated their own ineffective crisis management capabilities and national response policies supports such a statement.

Over the past twenty years, the number of migrants has been rising from 2.8 to 3.6% of the world population (IOM, 2020). The response of the governments of the destination countries has been small-scaled, that is, framed by attempts to change the policy of the government of the origin countries through bilateral and monoliteral agreements; actor-oriented, that is, overlooking the systemic root causes; and effect-target, that is, oriented to contain the own citizens' threat perception. Briefly, each destination country has adopted adaption and diversion responses creating the "new normal" of closing borders, integrating few, preferably skilled, migrants, and diverting flows toward other countries, the so-called externalization of the migration problem. A plan of mitigation and inactivation of the migration causes is not on the agenda of the governments. As a result, questions such as the following ones remain answered. Will the root causes of migration spontaneously deactivate? Are the countries of origin able to inactivate the causes by their own? Is the so-called aid to development, which has been the only response for the past seventy years, the adequate response?

A plurality of international organizations has been created to deal with migration. In 2006, the UN Global Migration Group was created to coordinate the UN agencies working in the field of migration. Today, it gathers 24 UN agencies. No doubt, UNHCR and the IOM have the primacy but with no coordinated strategies. To sending and receiving countries, such institutional proliferation is good for addressing migration through small-scaled, actor-oriented, and effect-targeted responses. As Rowan remarks, institutions that overlap in scope and membership increase the complexity of institutionalized cooperation and even raise the transaction costs that are supposed to lower unlike what happens with a single institution charged with addressing the collective problem (2021, 463). On such premise, two further questions are in order. Is the migration problem ripe to be addressed as a world-scale problem? Are there countries ready to act as the midwife of the multilateral policymaking process and bring the 2018 UN Global Compact on Migration to deliver?

The Preamble of the Compact expresses the wish of a "cooperative framework [that] uphold the sovereignty of states." By translating such sentence into a policy science sentence, the UN Compact wishes the formation of the world policy toward migration in respect of the principle of national ownership, that is through multilateral policymaking and consistent national policies. Briefly, to address the migration problem the apt way is to start a process like the climate change policymaking process bearing in mind that it will be a long and progressive process like the now thirty year-long process for building the world climate policy.

The UN should continue to be the midwife of such process but very much depends on the support of a good number of governments and the driving role of countries that have values and interest appropriate to the policy goals. The European Union is sitting in the back (Attinà, 2018). Only 19 member countries are of the group of countries that signed the Compact. In late 2021, the media reported that hundreds of undocumented migrants die every year, trying to cross the Mexico-US border. China leaders are silent on the migration problem. At the time of the world order transition, the relationship between these countries is crucial to address all world problems.

7 Conclusions

Multilateralism is the fit-for-purpose way to respond to world-scale problems through the sovereign states willingness to form both the world policies and the national convergent policies. It is also the legitimate way the more the decision-making rules give to each state the same decision-making rights, and the national ownership principle is respected. The present chapter has considered the lesson learned from the multilateral policymaking experience over the life cycle of the American world order. The strategic use of multilateralism by the coalition of the Western countries was decisive to re-building world order after WW II.

The subsequent delegitimization of the institutions and policies of the world order did not put multilateralism aside, but emphasized the need to stick to the principle of national ownership in multilateral policymaking. This is the most important lesson of the UNFCCC's climate policymaking process. Researching such process and comparing it with similar ones in project, such as the pandemic and migration multilateral policymaking projects, and in progress, such as the nuclear non-proliferation process, will widen knowledge about mastering the world-scale problems as well the future of order transition and world politics.

References

Abdenur, A. E. (2021). Climate action. Beyond the Paris Agreement. In T. G. Weiss & R. Wilkinson (Eds.), *Global governance futures* (pp. 257–272). Routledge.

Attinà, F. (2016). Migration drivers, the EU external migration policy and crisis management. *Romanian Journal of European Affairs, 16*(4), 15–31.

Attinà, F. (2018). Tackling the migrant wave: EU as a source and a manager of crisis. *Revista Espanola de Derecho Internacional., 70*(2), 49–70.

Attinà, F. (2021a). The world order lifecycle and world power competition. In F. Attinà (Ed.), *World order transition and the Atlantic area. Theoretical perspectives and empirical analysis* (pp. 13–28). Springer.

Attinà, F. (2021b). *World climate policy: Convergence and transition, ReShape* Paper no. 17, 2021, Jean Monnet Centre, University of Catania. http://www.dsps.unict.it/sites/default/files/files/ReShape paper 17_online.pdf.

Brown, G., & Susskind, D. (2020). International cooperation during the COVID-19 pandemic. *Oxford Review of Economic Policy, 36*(1), 64–S76.

Carr, M., & Lesniewska, F. (2020). Internet of Things, cybersecurity, and governing wicked problems: learning from climate change governance. *International Relations., 34*(3), 391–412.

Coleman, K. P. (2021). Slow progress on UN rapid deployment: The pitfalls of policy paradigms in international organizations. *International Studies Review, 23*(2), 455–483.

Crippa, M. (2020). *Fossil CO2 emissions of all world countries – 2020 Report*. Publications Office of the European Union.

Hafner, M., & Tagliapietra, S. (2020). *The geopolitics of the global energy transition*. Springer.

IOM. (2020). *World Migration Report,* Retrieved August 5, 2022, from https://worldmigrationreport.iom.int/wmr-2020-interactive/

Ivanova, M., & Escobar-Pemberthy, N. (2021). Biodiversity. Protecting the planetary web of life. In T. G. Weiss & R. Wilkinson (Eds.), *Global governance futures* (pp. 273–290). Routledge.

Kreienkamp, J., & Pegram, T. (2021). Governing complexity: Design principles for the governance of complex global catastrophic risks. *International Studies Review, 23*(3), 779–806.

Rowan, S. (2021). Does institutional proliferation undermine cooperation? Theory and evidence from climate change. *International Studies Quarterly, 65*(2), 461–475.

Ruggie, J. G. (1993). Multilateralism: The anatomy of an institution. In J. G. Ruggie (Ed.), *Multilateralism matters. The theory and praxis of an institutional form* (pp. 3–47). Columbia University Press.

Tausch, A. (2010). Globalisation and development: The relevance of classical "dependency" theory for the world today. *International Social Science Journal, 61*(2), 467–488.

Tausch, A. (2012). A globalization-oriented perspective on health, inequality, and socio-economic development. *International Journal of Health Planning and Management, 27*(1), 2–33.

Tausch, A. (2014). Reconsidering migration, globalization and social conditions in the world system. *Boğaziçi Journal, 27*(2), 63–82.

Tausch, A. (2015). Is globalization really good for public health? *International Journal of Health Planning and Management, 31*(4), 511–536.

Tausch, A., & Heshmati, A. (2011). Re-orient? Understanding contemporary shifts in the global political economy. *Journal of Globalization Studies, 3*(2), 89–128.

Tausch, A., & Heshmati, A. (2018). The effects of globalization on the environment. *UKH Journal of Social Sciences, 2*(1), 25–36.

Yang, H. (2021). Contesting legitimacy of global governance institutions: The case of the world health organization during the coronavirus pandemic. *International Studies Review, 23*(2), 113–134.

On the Hidden Pattern in the History of Wars

Tessaleno Devezas

Abstract Arno Tausch is undoubtedly a giant of the sociopolitical sciences, who contributed a lot to a modern and lucid vision of the world system that allows us today to better understand the succession of events that brought us to the current situation, which he very coherently calls as the Vortex of the Ukrainian crisis. He is one of the most prolific authors this author has ever known knew and with whom he had the opportunity to cooperate, and so he is honored to know that his publications may have in some way contributed to the maturing of his convictions. His publications are filled with tables and graphs exhibiting quantities, indexes, rates, etc., being very common the use of illustrated maps in which quantitative comparisons are highlighted. This essay presents an overview of the alluring relationship between K-waves and warfare, a field to which Arno Tausch has contributed significantly. Yet, in the light of this overview, it is presented an interpretation of the historical moment we are currently experiencing.

1 Introduction

Arno Tausch is undoubtedly a giant of the sociopolitical sciences, who contributed a lot to a modern and lucid vision of the world system that allows us today to better understand the succession of events that brought us to the current situation, which he very coherently calls as the Vortex of the Ukrainian crisis. One of the most prolific authors I ever know and with whom I have had the opportunity to cooperate, and I am also honored to know that my publications may have in some way contributed to the maturing of his convictions.

There is a distinctive aspect of his work, with respect to most other political and social scientists (who are very prone to purely verbal discourse), which concerns his rigorous quantitative approach, confronting theories with empirical data and using

T. Devezas (✉)
Atlântica – Instituto Universitário C-MAST (Center for Aerospace Science and Technologies), Lisbon, Portugal
e-mail: tdevezas@uatlantica.pt

G. Segell (ed.), *Development, Globalization, Global Values, and Security*,
https://doi.org/10.1007/978-3-031-24513-8_8

advanced statistical methods. His publications are filled with tables and graphs exhibiting quantities, indexes, rates, etc., being very common the use of illustrated maps in which quantitative comparisons are highlighted.

Very recently Tausch reactivated my memory regarding an event I organized in February 2005, when I was still active at the University of Beira Interior, in Covilhã. My objective, in this small article to be integrated in this volume dedicated to honoring the scientific journey of this extraordinary scientist, is to highlight some very interesting aspects that were discussed in this unique event.

2 A Rare Event

In the masterpiece opening this volume, Tausch refers to the above as a rare event of cooperation between NATO and the Russian Academy of Sciences, in which some gloomy predictions related to future warfare and turbulent times were analyzed and discussed. Indeed, it was a rare and fruitful event. It consisted in an ARW (Advanced Research Workshop) supported by the "NATO Program for Security Through Science," which promoted the gathering of about 41 scientists from all over the world (11 from USA, 12 from Russian Federation, 18 from Europe and other countries), and was entitled "Kondratieff Waves, Warfare, and World Security." The 12 participants from Russian Federation were mostly members of the relatively recently founded (1994) International Kondratieff Foundation, some of them members of the Russian Academy of Sciences. This scientific meeting resulted in the book mentioned by Tausch ("as a memorable book for our turbulent times"), published by IOS Press in 2006 (Devezas, 2006), as the volume 5m of the NATO collection E: Human and Societal Dynamics.

At that time, the major global concern was the fight against the growing wave of international terrorism, given that the attack on the Twin Towers (9/11) was still recent, the capture of Saddam had also taken place a short time ago (December 2003), Osama bin-Laden was still in an unknown whereabouts, Al Qaeda was a force to be reckoned with, not to mention that the Islamic jihad was on the rise. The workshop lasted for five days (February 14th to 18th, 2005) and was rich in presentations (38 in total) and discussions divided into three main topics: the Kondratieff Waves (Kondratieff waves revisited: new concepts on the interpretation of long-term fluctuations in economic growth), their relationship with the theme of wars and international security (Kondratieff waves and warfare), and finally conjectures about future developments in terms of economics and geopolitical conflicts (Looking into the future).

In my opening speech (also the Prologue of the book, titled Long Waves and Warfare: an Enduring Controversy) I noticed that there was a decreasing trend in the number of publications on Kondratieff waves (KWs for short), and I also noted the fact that some authors, who in previous years had published intensively on long socio-economic waves (LW for short), now had some doubts about their continuation, putting some hypotheses about a profound change in the world socio-economic system (for example, the globalization and intense digitization of the modern world) that could alter the cyclical and regular pattern observed until then. Some of these hypotheses were also presented and discussed during the workshop. Personally, I must confess that I too shared this possibility.

In the same speech I also pointed out the fact that many aspects related to the underlying driving forces of KWs remain unsolved, and that the most controversial one among these aspects is the relationship between KWs and the outbreak of major wars, first suggested by Kondratieff himself. This "enduring controversy" was the main motivation for this workshop, considering mainly the new emerging patterns of international conflicts, characterized by what is now called asymmetric warfare.

I concluded this foreword with a summary of the main discussions and conclusions that emerged from this workshop, highlighting the fact that extrapolation of the secular trend presented by some authors points to a severe conflict by the middle of the second decade of this century (upswing of the next KW), involving countries struggling for leadership in the Pacific Rim (Devezas, 2006, viii). In my own presentation during the workshop (Devezas, 2006, 245), titled the Emergence of Global Terrorism, I stated that "we were entering a very dangerous phase of warfare development, which may be controlled by the US and allies for a while, but can finally set the stage for a terrifying power conflict at the peak of the fifth Kondratieff wave" (2020s!).

When briefly discussing this NATO event of 2005, Tausch places special emphasis on the work of Goldstein, who in this ARW presented an update of his magnificent 1988 book (Goldstein, 1988). In his presentation, titled "The Predictive Power of Long Wave Theory, 1989–2004," Goldstein stated: "...The next phase, war, representing the dangerous second half of the long wave upswing, could then occur around 2015–2030, with the war peak possibly found in the late 2020s and the price peak around 2030 (50 years after 1980). In my 1988 book I pointed out to the period around the 2020s as a potential danger zone, and it still worries me" (Goldstein, 2006).

In his commentary about Goldstein conclusion presented above, Tausch adds: "Looking at the burning cities in Ukraine (April 2022), I am tempted to explain in despair: what a prophet he was!". Indeed, Goldstein's words were very prophetic, but we need to be careful with what was thought of at this time as a "danger zone," not only by Goldstein, but also by most ARW participants, or perhaps better said, by all experts and futurists active in the sociopolitical arena at the very beginning of the new century. With this purpose let us give a closer look at the other discussions that emerged during this unique NATO ARW.

Carefully reading the other chapters of the resulting IOS book (Devezas, 2006), we can easily observe that another conflict on European soil was unthinkable. The relatively recent conflict of the 1990s, with the involvement of NATO forces, which became known as the Balkan Wars or the Yugoslav Wars, was only the resolution of regional tensions of an ethnic nature, and independence wars, as well as a natural consequence of the USSR's disintegration and weakening of the communist system associated with an intense movement of Serbian nationalism, which resulted in the disintegration of the former Yugoslavia into a set of new independent countries, some of them are now NATO members (Croatia, Slovenia, Montenegro, North Macedonia). Resuming, this conflict was not a warfare conflict between powers struggling for global hegemony, the main cause of previous major conflicts generally occurring in the upswing phase of a KW (we turn to this point ahead).

We can shortly divide the main discussions arising during the ARW into three main strands: discussions on the very continuation of the K-waves (as exposed above, there were doubts about this), discussions on the relationship of KWs with major war outbreaks, and finally, forecasts and/or speculations about possible conflicts, as well as possibility of conflicts among hegemonic powers (or powers struggling for hegemony). Among all authors that presented disturbing predictions about warfare occurrences at the peak of the current wave (2020–2030), the prevailing thought was of serious conflicts with the new emerging power, China, challenging the hegemonic power at the top, the USA.

Alternatively, the unabated escalation of the Islamic jihad was also considered, which could spread through the Middle East and come to provoke a torrent of regional conflicts that could result in a confrontation of a global character. It is important to note that, among all the debates on future conflicts, at no time did any discussion about the resurgence of a cold war between the USA and Russia appear. Probably the most non-Russian participants judged the Russian Federation to be a card out of the deck when it came to hegemonic aspirations.

But this absence may reflect the typical myopia or cognitive biases of experts, who very rarely manage to extract from the complex mixture of trends emerging in the scenario of the world system, those that, although less likely to succeed, can emerge in force when synergistically interact with other hidden trends. Regarding this myopia, let us focus our magnifying glass on what my Russian colleagues were talking during the event. Before that, however, a brief observation is necessary: Mr. Vladimir Putin was a newcomer (since May 2000) on the world political scene, and he was undertaking profound reforms at the most diverse administrative and economic levels, sweeping away everything that was of less value from the previous Soviet system.

Certainly, our Russian peers were enthusiastic about the new directions taken by the new government still in the implementation phase. Many of them, when discussing Russian participation in the world scenario within the next decades, did so invariably predicting a very positive scenario. It is very curious to note that, in the chapter that closes the book, with the title "Cyclicity of Strategic Challenges in Russian History and Development Scenario for the twenty-first Century" (Ageev, 2006). In that Alexander Ageev, then director of the Institute for Economic Strategies, stated: "By the year 2030 under the «optimistic scenario» by academician A.G. Aganbegyan the GDP value of Russia will reach 4.5–5 trillion USD in current prices, that will make half of current value of GDP of the USA. By economic potential, GDP value, volume of industrial production, key assets and investments Russia will rise from the tenth to 4-fifth place and outgo Brazil, Italy, France, Great Britain, Germany and, probably, India, letting only the USA, China and Japan ahead."

The author proceeds mentioning that such optimistic scenario coincides with the scenarios advanced by other distinguished "academicians" like L. I. Abalkin, N.P Fedorenko, N. P. Laverov, and A. Y. Kontorovich, pointing out that "Russia will be able to return to the maximum level it had in the twentieth century before the years 2020–2030." It is important to note that the author cites these distinguished "academicians" without referring to their specific publications or reports. Moreover, he still adds that the structure of Russian economy will change dramatically and will be able to become one of the world leaders in no less than 8 to 50 macrotechnologies!

What an astonishing set of failed predictions, more like the prophecies of some frantic astrologer. None of this happened, in 2020 Russian GDP barely reached USD 1.5 trillion, remaining in tenth place in the global ranking. But not all was such bad in Ageev's predictions. At the conclusion of the chapter, the author presents a probabilistic estimate of the Russian "government transformation" for the next decades, resulting from the calculations of the experts of his team at the Institute of Economic Strategies, which demonstrates for 2020 the prevalence of majority influence (65%) of "one-party democracy," against 15% for effective democracy and less than 10% for authoritarianism (sic!). It so happens, however, that his logistic curves (substitution curves for forms of government) extrapolated to 2075 show a total decrease in authoritarianism and of the so-called one-party democracy, being replaced by some form of real democracy. We still have half a century ahead of us to wait for such a dream to happen.

3 Where Are We Now?

Tausch also poses this question in the opening chapter of this volume, and states that "world systems theory was, for decades, full of speculation about the future, and much of the world systems research writing projected a major global war around 2020 or 2030." This was also what could be observed in the presentations and emerged from discussions in 2005 during the NATO-ARW. But the dire question at the time was: a major war between (or among) which powers? And as already mentioned, the obvious candidates were the USA and China, along with other possible conflicts arising from the involvement of the Arab world, the Israel-Palestine issue, etc. But an interesting issue that I would like to focus our attention on in this short chapter concerns the exact temporal location of KWs, and their respective count. At the time of our ARW there was a consensus that we were in the transition trough between the fourth and fifth KW. However, in more recent times, a sixth KW has been pointed out as the wave currently under development. It is, in my view, a regrettable error, without empirical basis, which it is necessary to clarify.

Let us start with what was determined by Kondratieff himself. In his opus "The World Economy and its Conjunctures during and after the War," published in 1922, when he was only 30 years old (Kondratieff, 2004), he identified two and a half waves since the advent of the Industrial Revolution, with the following dating (his chap. V): an upward phase from 1789 to 1809 (peak in 1820), followed by a declining phase that inverted to upward in 1849, reaching a peak in 1873.

This peak was followed by another declining phase until 1896, when started the recent growing conjuncture (keep on mind that he was writing and using data until 1920). In the last sentence closing his deep analysis in this book we can read: ...“Therefore, once the propositions which have been brought forward and briefly motivated here prove true, the crisis of 1920, involved in a broader and more general prospect of conjunctures in time variation, seems not only a certain link in the minor capitalist cycle, but a highly remarkable stage in the succession of conjuncture waves of the long cycle. It is a starting point for a new and very original era of economic development and social relations.”

In other words, Kondratieff was claiming that the kind of mechanism that he had identified marking the rhythm of the succession of ups and downs of the world conjuncture since the IR, would continue to set the rhythm of the world socio-economic system (what he named as conjuncture) through the coming decades. He predicted the depression that started in 1929, which would later (after around 20 years, then a trough lasting until ~1950) be followed by a prosperous period of renewed growth.

As we know, this prediction of a renewed recovery of the capitalist system, which the Bolsheviks insisted on denying, forcing the belief that the depression of 1929 would be the certain end of the capitalist system, was what constituted the Kondratieff’s disgrace—a true damned prophet (killed in 1938 by Stalin, because he refused to withdraw his predictions about the cyclical behavior of the global conjuncture). Then, according to Kondratieff, we had KWs peaks in 1820 (first KW), 1870 (second KW), 1920 (third KW), corresponding to 2.5 waves, each with a mean duration of about a half century. Considering the unabated rhythm of the waves identified by Kondratieff, the next peaks would be expected to occur in 1970 (fourth KW) and 2020 (fifth KW) approximately.

In fact, the historical record showed that around 1970 we were reaching a growth peak, which ended with the famous oil shocks of the 1970s and culminated with the stock market crash of October 19, 1987 (Black Monday), thus starting a long period of recession as predicted by Kondratieff. The culmination of the fourth KW was thus drawn, and on this occasion the theory of “long waves of the world conjuncture” was rediscovered and an avalanche of publications on the phenomenon broke out in the 1980s and the 1990s. Among these appeared a couple of books (Mager, 1987; Tylecote, 1993; Prechter, 1996) bringing to light a very interesting graph published in 1974 by the now-defunct Media General Financial Services, which was also reproduced in my 2001 paper (Devezas & Corredine, 2001), here exhibited in Fig. 1.

What is amazing about this graph is that having been published in 1974, it predicted in detail the further unfolding of the fourth KW (dotted line), even anticipating a serious secondary depression in the late 1980s, that is, anticipating the stock market crash of 1987. Consequently, soon after the “black Monday” the graph appeared also in the pages of “The Economist.” The events following the outbreak of the recession in 1987, continuing into the early years of the twenty-first century, are very well described by the dotted line extrapolation.

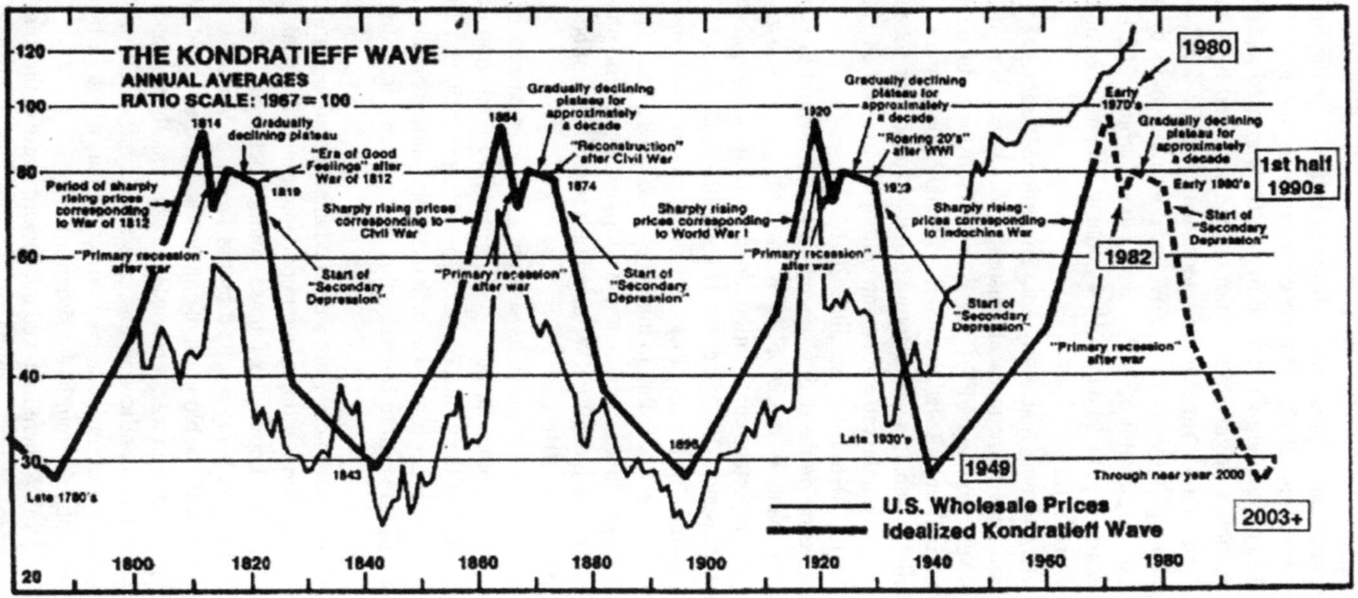

Fig. 1 Graph of four Kondratieff waves originally published by Media General Financial Services in 1974 and reproduced in some books in the 1980s and the 1990s

Important to point out that this graph was used also in the cover of two books of the book series (Almanac) "Kondratieff Waves" (Grinin et al., 2012, 2016), the same books in which some authors presented a different scheme for the timing of the waves (Korotayev & Grinin, 2012). This new scheme was first introduced by the German economist Leo Neofiodow (Neofiodow, 1996), who used a skewed dating system from Kondratieff's original publications, soon followed by some Russian authors who in the mentioned Almanac misinterpreted the 2007–2008 crisis as the final descending phase of a far shorter fifth KW, and not a momentary disturbance of the ascending phase of the fifth KW that would peak around 2020–2030, as I demonstrated in my article published in 2010 and that was reproduced in the referred 2012 volume (Devezas, 2012).

It is very important to note that in my above-mentioned articles, I was able to empirically demonstrate the unequivocal existence of a succession of four long socio-economic waves, applying FFT (Fast Fourier Transform) to time series of important economic indicators (Maddison series of GDP, Dow Jones, purchasing power Gold/US dollar). Also, it is worth noting that the same dating scheme as mine was used by other authors in the same Alamanc volumes (Modelski, 2012; Helenius, 2012; Thompson, 2012, 2016; Berry & Elliot, 2016).

Closing this discussion on the dating scheme of KWs, I would like to highlight the fact that all participants in the 2005 ARW used the system shown in Fig. 1, including Goldstein's important contribution already mentioned (Goldstein, 2006). See also the graph presented by Marchetti (Marchetti, 2006, Fig. 1). Goldstein developed an interesting "war time clock" for analyzing warfare, distinguishing peak wars and trough wars. According to this scheme, 9/11 was one of those trough conflicts that in his words "may signal a new Kondratieff expansion phase," therefore, the starting point for the fifth KW, and not a sixth as Neofiodow and his followers want (Neofiodow, 2016).

As already mentioned in the previous section, some participants of the 2005 ARW, including me and Goldstein, put a high probability prediction of a violent armed conflict at the peak of the fifth KW, sometime between 2020 and 2030, which Goldstein qualified as a "danger zone." Unfortunately, the current circumstances and the recent historical record seem to reinforce this idea, which I try to reproduce in Fig. 2 below.

I do not wish in this paper to carry out a detailed analysis of the unfolding of the previous waves, but it is worth noting some important points. The peak of the third wave characterized, as we know, the intense conflict between European superpowers, which ended up not being resolved by the Treaty of Versailles (1919) and remained to be resolved with the bloodiest of conflicts so far, the second WW, which had the characteristic of a trough war. It consisted fundamentally of a macro decision phase to resolve unresolved conflicts, in this case the dispute in Europe, to which was added the interference of an incumbent power, Japan, which in fact gave the conflict a truly global nature (a multipolar situation).

Soon after the second WW there followed the upward phase of the fourth KW, the peak of which characterized the warming of the cold war (a dispute for world hegemony). Then followed the descending phase of the fourth KW, when we had

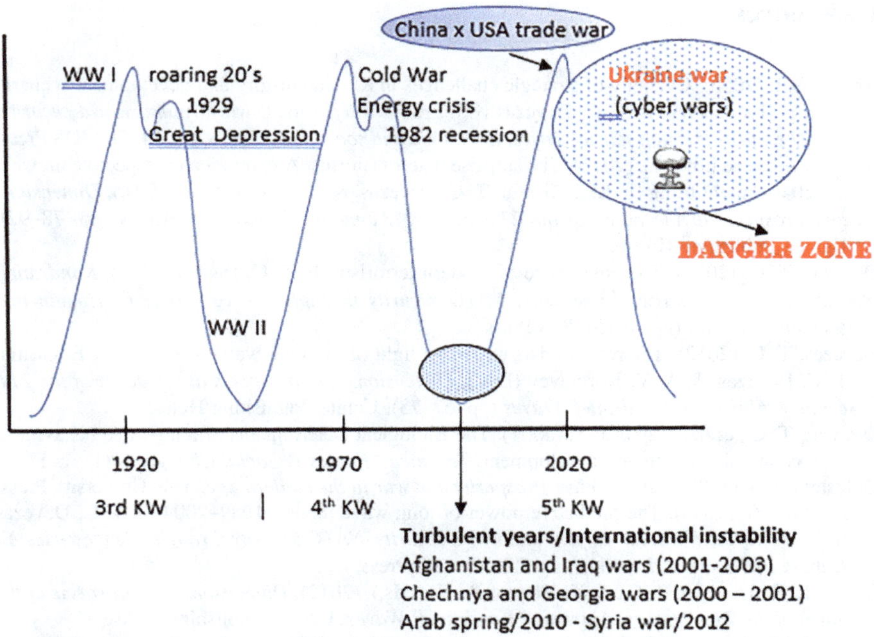

Fig. 2 Succession of the third, fourth, and fifth Kondratieff waves, highlighting the important events occurred during their unfolding, with emphasis to the events during the last 20 years

typical cases of economic/political downturns, as the collapse of URSS and the Balkans war, soon succeeded by the turbulence of trough wars, as indicated by the arrow under the graph. The decade from 2000–2010 saw conflicts in various parts of the world but the most violent occurred in Afghanistan, Iraq, Chad, and Chechnya, as well as in Syria (2012).

After that we had the typical occurrences when approaching the peak of KWs, this time the struggle for hegemony opposing USA and China. Now Russia entered the stage as an incumbent power. It is worrying that, even if the current war in Ukraine ends in the next months, the problems underlying its causes will remain to be solved and could even get worse. Once again, a highly dangerous situation of multipolarity emerges, that is, a framework of unresolved issues that, as on previous occasions, could have dire consequences. Further speculations do not contribute much to the timid intent of this brief chapter.

My final word is addressed to my friend Arno Tausch. This complex scenario requires a profound revision of the known underlying mechanisms driving political conflicts. Perhaps, with his deep knowledge of the workings of the global political system, he can show us if there is a chance of get out of this conundrum.

References

Ageev, A. I. (2006). Cyclicity of strategic challenges in Russian history and development scenario for the 21st century. In T. C. Devezas (Ed.), *Kondratieff waves, warfare and world security, NATO security through science series, E: Human and societal dynamics* (pp. 34–56). IOS Press.

Berry, B. J. L., & Elliot, E. (2016). The surprise that transforms. An American perspective on what the 2040s might bring. In L. E. Grinin, T. C. Devezas, & A. V. Korotayev (Eds.), *Dimensions and prospects at the dawn of the 21st century, Alamanac Kondratieff Waves* (pp. 78–92). Uchitel Publishing House.

Devezas, T. C. (2006). The emergence of modern terrorism. In T. C. Devezas (Ed.), *Kondratieff waves, warfare and world security, NATO security through science series, E: Human and societal dynamics* (pp. 1–15). IOS Press.

Devezas, T. C. (2012). The recent crisis under the light of the long wave theory. In L. E. Grinin, T. C. Devezas, & A. V. Korotayev (Eds.), *Dimensions and prospects at the dawn of the 21st century, Alamanac Kondratieff Waves* (pp. 67–78). Uchitel Publishing House.

Devezas, T. C., & Corredine, J. T. (2001). The Biological Determinants of long-wave behavior in socioeconomic growth and development. *Technical, Forces & Social Change, 68*(1), 1–57.

Goldstein, J. S. (1988). *Long cycles: Prosperity and war in the modern age*. Yale University Press.

Goldstein, J. S. (2006). The predictive power of long wave theory, 1989–2004. In T. C. Devezas (Ed.), *Kondratieff waves, warfare and world security, NATO security through science series, E: Human and societal dynamics* (pp. 20–32). IOS Press.

Grinin, L. E., Devezas, T. C., & Korotayev, A. V. (Eds.). (2012). *Dimensions and prospects at the dawn of the 21st century, Alamanac Kondratieff Waves*. Uchitel Publishing House.

Grinin, L. E., Devezas, T. C., & Korotayev, A. V. (Eds.). (2016). *Cycles, crises, and forecasts, Alamanac Kondratieff Waves*. Uchitel Publishing House.

Helenius, A. (2012). Waves on waves: Long waves on the seven seas. In L. E. Grinin, T. C. Devezas, & A. V. Korotayev (Eds.), *Cycles, crises, and forecasts, Alamanac Kondratieff Waves* (pp. 130–145). Uchitel Publishing House.

Kondratieff, N. D. (2004). *The world economy and its conjunctures during and after the war*. International Kondratieff Foundation.

Korotayev, A. V., & Grinin, E. L. (2012). Kondratieff waves in the world systems perspective. In L. E. Grinin, T. C. Devezas, & A. V. Korotayev (Eds.), *Dimensions and prospects at the dawn of the 21st century, Alamanac Kondratieff Waves* (pp. 78–86). Uchitel Publishing House.

Mager, N. H. (1987). *The Kondratieff Waves*. Praeger Publishers.

Marchetti, C. (2006). Is history automatic and are wars a la Carte? The perplexing suggestion of a system analysis of historical time series. In T. C. Devezas (Ed.), *Kondratieff waves, warfare and world security, NATO security through science series, E: Human and societal dynamics* (pp. 41–58). IOS Press.

Modelski, G. (2012). Kondratieff (K-) waves in the modern world system. In L. E. Grinin, T. C. Devezas, & A. V. Korotayev (Eds.), *Dimensions and prospects at the dawn of the 21st century, Alamanac Kondratieff Waves* (pp. 134–145). Uchitel Publishing House.

Neofiodow, L. A. (1996). *Der Sechste Kondratieff, Wege zur Produktivität Voll Beschäftigung im Zitalter der Information*. Rhein-Sieg Verlag.

Neofiodow, L. (2016). The Sixth Kondratieff – The new longwave of the global economy. In L. E. Grinin, T. C. Devezas, & A. V. Korotayev (Eds.), *Dimensions and prospects at the dawn of the 21st century, Alamanac Kondratieff Waves* (pp. 152–169). Uchitel Publishing House.

Prechter, R. R., Jr. (1996). *At the crest of the tidal wave*. New Classical Library.

Thompson, W. R. (2012). Energy, K-waves, leading economies, and their interpretation/implications. In L. E. Grinin, T. C. Devezas, & A. V. Korotayev (Eds.), *Dimensions and prospects at the dawn of the 21st century, Alamanac Kondratieff Waves* (pp. 155–168). Uchitel Publishing House.

Thompson, W. R. (2016). Revisiting a long-term perspective on Kondratieff phenomena. In L. E. Grinin, T. C. Devezas, & A. V. Korotayev (Eds.), *Dimensions and prospects at the dawn of the 21st century, Alamanac Kondratieff Waves* (pp. 169–183). Uchitel Publishing House.

Tylecote, A. (1993). *The long wave in the world economy: The present crisis in historical perspective*. Routledge.

Russia and Germany in the System of Powers

Hans-Heinrich Nolte

Abstract This essay examines the full circle in the context of the war between Russia and Ukraine that started in 2022 in that Arno Tausch convincingly coined the basic relation between the European World System and Russia as treadmill—always laboring to compete with the West, maybe successful in the military (Suvorov crossing the Alps, Russian troops in Paris and Berlin…) but never quite reaching its economic and social standards.

1 Method: World-System-Studies

My understanding of history is indebted to the attempt to integrate religious, social, economic, political, and other fields in social sciences. As a model for that attempt, I use the World-System approach. In methods to my judgment this leads to nonlinear research—not only for those periods, when economic or social data anyway are lacking. Using other, for instance religious sources (Nolte, 2015), does change not only our basis in sources though, but also our historiographic setup, since before trying integration we have to grasp quite numerous and different narratives—from the history of globalization to that of a Church for instance, from military history to that of an intellectual interchange—for instance, Chaadaev's lamentation about the loneliness of Russia to Herzens hope for a new and fresh people entering the politics of Europe (which he saw as aging). Both were united in critic of the Russian Empire, which (as Herzen saw it) cut the Russians off from humanity laboring to reach more freedom described by Herzen (1969).

The narrative Russia and Germany at a first glance does seem to be a simple one. But on second glance it does offer a couple of specific difficulties. What kingdom or Empire may we identify as Germany or Russia in what period? When did modern nations develop from Empires (Nolte, 2017)? Where did the Netherlands belong and where Belorussia, what about the kingdom of Bohemia or the city of Novgorod the

H.-H. Nolte (✉)
Historisches Seminar der Leibniz-Universität Hannover, Hannover, Germany

G. Segell (ed.), *Development, Globalization, Global Values, and Security*,
https://doi.org/10.1007/978-3-031-24513-8_9

Great? (Stökl, 1967; Wernstedt, 2018) From what date on did some people speaking German identify themselves as Austrians or some people speaking Russian identify themselves as Ukrainians? In 2001, 11.5% of the Ukrainian population considered themselves as Ukrainians of Russian tongue, 17.3% as Russians. These two groups constituted 38% of the population in the southern and 52% in the eastern Ukraine as analyzed by von Werth (2007, 5–9).

This sketch can, of course, not represent the broad and multicolored picture and indeed skips such a central topic as East-Middle-Europe between Russia and Germany, but may, I hope, add the brush of a special relation.

2 Medieval Settings

Both Russia (Nolte, 2012) and Germany (Leuschner, 1985) as political entities started in the ninth century, but while the first was Christianized in obedience to Constantinople, the second in obedience to Rome. The first form of Christianity may be characterized as more philosophical and the second as more juridical. While Russia developed in the political tradition of the Patriarch accepting second place behind the Emperor of the Eastern Roman Empire (which we call Byzantine), Germany (the kingdom Germania) following the Carolingians inherited the tradition of Western Roman Empire. Having defeated the nomadic and heathen Hungarians in 955 Otto 1st attempted the renovation of the Roman Empire. It expanded to the West into Burgundy, Lorraine, and northern Italy and to the East into Bohemia and Silesia. This multinational state was dependent on military victories, it did not even have a continuous capital and was characterized by a feeble bureaucracy, dependent in fact on the Catholic Church.

The Schism between East (Orthodoxy) and West (Catholicism) in 1054 legitimized the expansions of West-Roman Christianity into the East, highlighted by the 4th crusade in 1204, when the Orthodox capital Constantinople was plundered, and the Empire divided between the victors. The 4th crusade was seen as a betrayal in the Orthodox world, and when in 1240–1242 the Teutonic Order and Sweden attacked Russia, which had just been subjugated by the Mongols, Alexander Nevskij became the hero of defense against the West (Schenk, 2004).

Russia's strategic place in Christianity was different—continuously it had to fight with nomads, was conquered 1237ff., and up to 1480 survived as a peripheral province of the Mongol Empire, while Sweden expanded to Finland and the Teutonic Order conquered the southern coast of the Baltic. A territory stretching from Gdansk (today Poland) via Kaliningrad (today Russia) and Klaipeda (today Lithuania) toward the modern republics of Latvia and Estonia.

The Mongol Empire also enabled Lithuanian expansion and following the marriage of Jagiello and Jadviga in 1386 the expansion of Poland/Lithuania across the Dnjepr till Smolensk. The Ottoman conquest of Constantinople in 1453 further isolated Russia, and during the Counterreformation (1596) the Orthodox bishops in the Polish Ukraine and Belorussia joined a Clerical Union with Rome (Donecker, 2008).

Germany between 955 and 1529 was not endangered by nomadic powers, since Hungary following its defeat turned into a Roman-Christian kingdom. When the Emperors lost the fight with the Catholic Church in the twelfth and thirteenth centuries—the last one, Fredric II was more an Italian or rather Sicilian than a German, by the way—the regional lords (domini terrae) gained power. When elected as king of Germania he almost did not know the German language (Kantorowicz, 1987, 32; Leuschner, 1985).

The Teutonic Order—founded mostly by low German nobility—for the territories on the Baltic accepted overlordship from Rome, not the Empire. There for more than a century the biggest one, the kingdom of Bohemia was leading (and the fundamental law Golden Bulle in 1356 prescribed that the sons of the electoral Princes had to learn Czech). But in 1526 the rise of the German language Habsburgs started, and in 1620 the Bohemian nobility was defeated and dispossessed.

3 Germany and Russia in the European World System

Following the final defeat of the numerous emperors in the thirteenth century Western Europe was organized as a system of kingdoms, kept together by Catholic institutions, especially in law, science, culture, and of course religion. The internal relations were characterized by competition, fight for hegemony, territorial expansion and accumulation of competences, especially military ones (Hoffmann, 2015). The West boomed not only in military, but also in the economy. A regional division of labor was developed. The center, still today the European banana stretching from Flanders to Firence, accumulated wealth by trade and traffic, exporting money (silver), selling colonial goods as Chinese porcelain and Indian cloth, some industrial products as weapons and linen. The semi-periphery in Spain, Sweden, or Russia exported raw materials like wood or grain and imported money (silver), colonial goods like sugar and industrial goods as weapons and linen. Also, in the Baltic trade increased. In the seventeenth century the Dutch had taken over from the German Hanse, and in the eighteenth the British took over from the Dutch (Nolte, 2005).

Arno Tausch (1991) convincingly has coined the basic relation between the European World-System and Russia as treadmill—always laboring to compete with the Center, which seen from Russia was and is the West. Successful, in the military (Suvorov crossing the Alps, Russian troops in Paris and Berlin. . .) but never quite reaching its economic and social standards. In Russia anyway the attempts to reach Western standards had many successes, especially in the military, but developments never were fast enough to really catch up.

In the Russian opposition this acceleration was criticized, already in an Ukaz of Peter III, respectively, Emel'jan Pugachev in 1774, promising that when all enemies and the noble villains are extirpated everyone may expect quiet and a peaceful life in eternity. If we put the hatred of Cossacks against nobility and bureaucracy aside—looking to the East for quiet and normal life, for simple relations between persons, for less complexity may even remind of Goethe's hope for patriarchs in the East.

But back to the system. As noted, it expanded in the West. Based upon the 1st Industrial Revolution (ships and mechanics) the Americas were conquered; the Ottoman Empire was bypassed and a direct connection to the old Sino-Indian trade system established. On the sea European navies defeated all others. They fought one another—Portugal and Spain, the Netherlands and Great Britain—but no Asian Power was able to stand up to them on the seas.

On land though, following Tamerlans organization of weapons-manufactures, logistics, and artillery, Asian Empires won ground-battles. The Manjou stopped Russia's advance toward the Amur and the Ottomans arrived at the doorsteps of Kiev in 1672 and those of Vienna in 1683. During the seventeenth century though European powers invented another military Revolution—how to win battles by continuous gunfire and build fortresses that were able to offer resistance. And they enforced strict discipline in their military and more than that, in their societies (Darwin, 2008).

Following its growth of power and liberation from the Mongols in 1480 in Muscovy Church, nobility and merchants took part in politics. Ivan IV made an unsuccessful attempt at absolutism, but in his time the course of the Volga-River toward Astrakhan successfully was conquered and that of Siberia started (Nolte, 2016).

Ivan IV also tried to gain the Baltic coast, but Moscow was defeated and now really cut off from the Baltic Sea by Sweden. Poland attempted to annex Russia but was turned back by a Romanov-restauration of Tsardom. Muscovy mostly kept peace with the Ottoman Empire, but when the Ottoman resistance in Hungary and Podolia proved enduring, the Tsar was invited to become an ally of the Pope, Austria and Poland in 1686. The Ukraine was divided between Poland and Russia along the Dnepr River.

In Russia, the seventeenth century was one of reforms adopting West-European standards to Orthodox Russian preconditions. Old export-products as fur were multiplied following the expansion toward the Pacific, and agrarian products like hemp, linen, and tar, but also semi-manufactured goods like ropes were added. In economics Russia stayed part of the semi-periphery—importing silver, manufactured goods, and specialists. In politics though Russia was on the way toward a considerable power—it organized the basics of the military revolution.

Modern ironworks, introduced by supporting a German entrepreneur, started competition with old production-sites, infantry and artillery were enlarged. More taxes and services of peasants were enforced, and Russia was the last country in the Baltic region, where peasants lost the right to choose their feudal lord and were "written to the land" (Schmidt, 1997; Smolarz & Nolte, 2023). In uprisings of the internal peripheries, called peasant wars in Soviet times, some realized that foreign specialists were active in these innovations adopting western models to Russian conditions but also at infringing the liberties of the common men (which in Russia were called black people).

Muscovy had taken part in European expansion toward China and the Caucasus. Peter Ist realized the importance of a navy for a modern power, and when despite the alliance with Austria the Black Sea remained closed for Russian shipping he allied with Poland and attacked Sweden, conquered the Baltic, and founded a new capital on the shores of the sea—window to Europe, as Pushkin has called it. In 1721 Russia graded itself up as Empire.

Germany since the reformation was divided between Protestantism and Catholicism fought both against the hegemonial expansion of France in the West and the Ottoman Empire in the East. It was happy if it kept the ground (Strasburg, for instance, was lost in 1681). Germans only occasionally took part in European expansion, as migrant laborers in Dutch farms and ships or as soldiers and specialists in Russian service.

While Germans were mocked as *moffen* in the Netherlands in the Russian Empire, they were able to make careers and even become ennobled by the Russian table of ranks. When in the eighteenth century Austria, Prussia, and Russia carved up Poland between themselves, in all three absolutist states German bureaucrats and soldiers served and settlers were invited also into those parts of the Ottoman Empire conquered by the Russian Empire. The relationship was asymmetrical in a double way: while Russia since Peter Ist was a big power, German powers battled against one another were conquered by France in Napoleons Time and only regained independence by British resilience and Russian campaigns till 1815 (Stökl, 1967).

Many Germans, generally alphabetized in comprehensive schooling or studied in one of the many small German universities made careers in Russia—where they received better pay than Russians in the same jobs. A difference in habitus became important: Germans usually had learned discipline under the rule of small-state Polizei and in the course of Konfessionalisierung, while in Russia this could not be accomplished, despite attempts in the reign of Peter Ist (Donecker, 2008).

4 Imperialism

Such competitional cooperation went on in the nineteenth century, when Great Britain based on the 2nd industrial revolution (Steam-power and machines), the global navy, and the defeat of Napoleon became the hegemon of the System. Armies of the East-India Company conquered India and following the defeat of a national uprising called their colony Empire 1877 which name then was transferred to the whole complex of conquered territories from Ireland to Cape Town. Great Britain even reached out for China. Austria had taken the name of Empire in 1804, shortly before Napoleon founded the French one.

In France, the 2nd Empire was founded in 1866, but Prussia—most successful in competing imitation of Britain—defeated Austria and France and founded a new German Empire, the 2nd in case we consider the Holy Roman Empire as German. The identity German excluding Austria remained uncertain, though—on both sides of the border, for instance the Austrian Hitler served in the Bavarian Army during the 1st World War and became legally German only in 1932. Based upon the 3rd Industrial Revolution (Electricity and Chemics) Germany started an own attempt at hegemony, looking for own colonies.

Since everyone who had some power called it an Empire now, the period convincingly is termed Imperialism. With pomp and circumstance the monarchs tried to win the acclamation of their peoples. Both German and Russian intellectuals read Kant and Hegel and disputed about the historical roles of their nations. Urbanization grew, working people flocking into the growing cities, and parliamentarism was fought for with some success—in Germany since 1871, in Russia since 1907.

Especially German middle classes followed the military habitus of the nobility—in the role of the Prussian lieutenant of the reserve. In the Russian Empire a more outspoken part of the *Intelligencia* was against it and their image of the military was more characterized by the Cossacks. In both societies the national movements won votes—Poles, for instance, in all three parts of the old Poland. The Social Democrats, both with Marxist programs, won support, but in Germany in the debate on Bernstein many hoped for a parliamentary and not revolutionary way to socialism. In Russia, the Social-revolutionary Party was strong, which had no counterpart in Germany, while the idea of archaic peasant communities in Russia fascinated many not only on the left, and yet more Germans looked to Russia for a simple life—Rilke went on a pilgrimage to Tolstoy.

Why did the Russian Empire not intervene against the rise of Prussia? France and Britain championed liberalism and humiliated Russia in the Crimean War, and Prussia showed a path of more conservative and yet militarily successful development in Europe. Economically the Russian Empire boomed before the 1st World War—without being able though to provide sufficient industrial work for the peasants forced to look for it, since their plots were too small as examined by Kahan (1989). Again, as since the times of Peter 1st, Armament industries formed a central part. The Russian population outgrew the German one by far, the economic potential was developed by English and Belgian capital, sponsored by Petersburg, and Russia expanded its colonies in Central Asia, checked only by Japan 1905.

Germany in colonial politics was a latecomer (the best pieces of the overseas peripheries already had been taken, mostly by Britain and its settlement-colonies) and because of simple geography (the German ports are situated in the inner reach of a small coastal sea, not the Atlantic itself) in the 1st World War, Germany was blockaded in the West, also in foodstuffs. German victories at the Eastern Front though, following Tannenberg seemed to point into the other direction, and the Treaty of Brest-Litovsk 1918 for some months opened the possibilities of German colonies in the East, where no British Fleet was able to intervene. Only the defeat in the West stopped this expansion toward Ukraine and Caucasus. Victories in the East did not decide the war though—that was done by the entry of the USA.

5 Germany and Soviet Union: 1918–1945

The German Republic of 1918 and the SU as a new state attempting to leave the system of capitalist powers and start a socialist world of solidarity cooperated following the treaty of Rapallo. Following the World Economic Crisis, when western powers used their colonial Empires overseas to stabilize their staggering economies, the German Republic even prefinanced industrial exports to the SU with the aim to bring the growing numbers of unemployed down. But in 1933 the National-Socialists started an anticommunist dictatorship, offering Jews as first victims for the enrichment (*Arisierung* of Jewish property) of its followers in the SA (*Sturmabteilungen*).

The concept of colonial Empire in the East inspired the sergeant Adolf Hitler, who in his first book Mein Kampf (written while in jail following his Putsch) thought up an historical chance for Germany to become a World-Power, figuring that Jewish bolshevism had destroyed the old Germanic Russian elite and by that opened Eastern Europe for German expansion, since Jews—as he thought—while able to disintegrate a state were not able to govern one (Hitler, 1935). This adaptation of Gobineau's verdict on the French revolution and the fake-news emigrants from Russia created about the Jewish character of bolshevism, was directed against Weimar foreign politics of raison d'etat cooperation with the Soviet Union, often sketched as Rapallo.

This racist miscalculation was disproved by the political and industrial buildup led by the Communist Party (in which Jews as an urban ethnical group of course were more represented than other ethnic groups). Politically nations and nationalities within the territories held by Bolsheviks at the end of the revolutionary wars in 1921 were organized in own territories, the bigger ones as Socialist Soviet Republic (SSR) and united in the Union as USSR, which meant that in international politics the term Russia disappeared. Economically the Communist Party ended the commune-organizations of the Russian peasantry and started a program of forced industrialization by central planning, using state-terror for political power, the budget and confiscations for capital, and forced labor for work (Fitzpatrick, 1999).

In the 3rd Five-Year Plan the SU developed the industrial base between Oural and Siberia, difficult to reach from the West. It counted in 1940, 194 million inhabitants producing coal, steel, and oil—outproducing the Germans in the Reich in the same year in many fields. And most of all: in Germany since the war 16–17 persons were born on every 1000, while in the SU the ratio was 30–40, resulting in the fact that much more young men could be called to arms in the East—despite the extreme losses, which the campaigns against bourgeois, anticommunists, or communist deviators and especially Stalinist agrarian politics had caused (Kennedy, 1989).

Hitler's intellectual construct (mis-)guided German hope to defeat and destroy the SU fast and start politics of economic exploitation and settler-colonialism. Despite Soviet superiority in manpower and arms Germany—battling Great Britain in the West—attacked the SU in 1941 and, despite some stunning successes in the first months, lost the battle for Moscow the same year. With that battle it lost the war since that now changed into a war of attrition. Already afraid of defeat the Nazis burnt all bridges, which might have been taken back to culture, and committed systematic genocide against Jews, Sinti and Roma and also genocidal starvation and massacres against Soviet POWs, cities like Leningrad and regions with partisans. Kay (2021) estimates 8.5 million civilians murdered in the Soviet Union. About half of the six million murdered Jews were killed there. Overall the number of dead or unborn Soviets is estimated at about 27 million.

6 Germany and the Soviet Union in the Cold War

As partner in victory the SU following 1945 was a World-Power, globally recognized as able to veto the decisions of the newly established UN-Security-Council, soon the second atomic power and maybe even the first one able to deliver their bombs by intercontinental rockets. The SU imposed on almost all territories liberated by the Red Army from German power socio-economic systems following the Soviet model, even in Poland—where the percentage of Communists in the population never was big—and in the Soviet zone of occupation—where the foundation of the German Democratic Republic split the German nation and even the capital Berlin (Schlögel, 2017).

Polish critical communists called the new political system monopoly-socialism—the socio-economic power of the capitalists was broken, that did not result in more democracy though, but in a bureaucratic monopoly of the leading officials of the Communist Parties (respectively, the Parties united with Social-Democrats by force) as examined by Modzelewski (1969). But the planned economies were not able to outproduce the capitalist ones, and the improvement of living-conditions in the West was faster than in the East.

There were a couple of reasons the economic shortcomings—the losses of millions of young people in the War, the problems of follow-up industrialization, the militaristic concept, that keeping up with the West was especially important in the arms-race, and mistakes in planning. But the overall effect was disastrous. Especially people in working-age emigrated en masse—and when walls and barbed wires were considered necessary to stop such exodus, that already signaled defeat in the main question: what system served people better? But the SU had thrown in its power on the side of the German Communists, representing maybe 10% of the East-German population.

While the Soviet Union had problems in the economy, West-Germany as member first of the Montan-Union and later of the EU experienced a fundamental *hausse*, which changed the country thoroughly. In one regard Germany became more like Russia, where Muslims have lived since Medieval times: For the first time in German history, a sizeable Muslim minority migrated toward the Rhine river.

In politics, while the SU was a global power, the Federal German Republic might be called third-rate with a growing economic and financial capability, but without own disposal of troops (rearmament created only troops under NATO-command). Anyway, in the last decades of the Cold War, German Social-democrats with Willy Brandt as chancellor developed the concept of change by getting closer. Without being able to decide whether politics along this line or American acceleration of the armament-race by introducing Star-Wars brought about the decision for a rebuilding of socialism including market-elements (perestroika) reading Gorbachev anyway shows some influence from German politics.

7 End of Soviet Union and Reunification of Germany

The data to my understanding offer a clear picture. The USA without any doubt is the most powerful country in the world, with the biggest percentage of very rich people (who combine economic and political influence) and the biggest military establishment. China may be on the way to the top. It is reasonable to consider the USA and China the first-rate powers of the beginning twenty-first century (Table 1, Nolte, 2018).

Of the five powers compared, Russia and the UK may be considered second-rate—in BIP per country they are far below the first two, but again there are rich people especially in the UK and considerable military establishments. Germany, the only of the five without Veto and Atom bombs, may be seen as a third-rate power, with relatively little military expenditure and the most balanced social structure if we follow the GINI-Index.

Of course, there are many other indicators, as for instance the percentage of people in jail per country. One indicator anyway should be considered, since parliamentary systems on the long run are more successful than authoritarian ones: USA and UK have a long tradition in parliaments, China and Russia have little, and Germany has a broken one.

8 Russia

The reasons for the end of the SU are discussed in other places. Following the dissolution, the former Socialist Republics organized themselves as capitalist democratic nation-states, although in most (excepting the Baltic ones) there were neither established republican traditions nor enough politically independent people, to offer favorable conditions for democracy. A fundamental difficulty was the fact that big properties were largely not legitimized. In Russia there had been no impressing leaders of industry, traffic or banking like Krupp, Ballin, or Bleichröder and privatization of Soviet property was enforced by a disputed system of vouchers creating a new social group: the Oligarchs. While Russian BIP was decimated and almost a

Table 1 Formal comparison. A table may help in the comparison

	1	2	3	4	5	6	7	8	9
USA	4.6	56	18.6	37	12	x	x	610	3.5
China	20.3	8.2	11	50	<1	x	x	216	2.1
Russia	3.0	9.8	1.3	36	2	x	x	84.5	4.5
Germany	1.3	44	3.1	28	2	x	–	46.7	1.2
UK	1.0	42	2.4	35	6	x	x	60.5	2.2

1. percentage of world-population, 2. BIP per head 2016, 3. BIP, 4. GINI 2012, 5. Percentage of population belonging to globally richest 1%, 6. Cyber-capacity, 7. Veto, 8. Billion $ Military 2014, 9. Percentage of BIP

quarter of the population fell below the poverty line, these Oligarchs between 1992 and 1980 exported some USD $150 billion to the world market, establishing a new sector of the global rich.

When Putin came to power in 2000 liberalism in Russia had lost public trust. The political system has been characterized as kind of a dual state: a democratic, federal system on the one side and a system of patronage as a main way of power running from the president (Sakwa, 2010) on the other. The patronage-system in Russia in fact is running down to the kiosks—even they need a roof (Varese, 2001).

Putin started with a period of experimenting whether Russia would find acknowledgment as a world-power in the continuation of the SU. In the memorandum of Budapest, Russia guaranteed the borders of the Ukraine in exchange for monopolizing formerly Soviet atom-weapons and receiving Veto-power in the UN. Many considered this more cooperative period of Russian foreign politics as simulation.

Indeed, the Socialist Union reviewed from a nation-state had some similarities with an Empire, and there were postimperial traumata. I consider it as a period of finding a role in the global system. There a triumphalist mood was predominating; many believed that it was possible to solve the global problems with neoliberal rules. For many Russians, these rules seemed partial though—while the West considered his own activities in Afghanistan as liberation, it saw the Chechen-wars as subjugation.

In the following years NATO intervened in the dissolution-wars of Yugoslavia militarily without consent of the Security-council. Convincingly condemning the Serbian massacre as genocidal NATO took the fake news that there was a Serbian horseshoe-plan to murder Albanians as reason for military actions, which in the end facilitated the independence of the province of Kossovo. Most important was the attack of the USA and the coalition of the willing against Iraq relying on the fake news, that the Iraq was producing weapons of mass-destruction. USA and Britain prevented a judgment of Security-Council on their break of the UN-Covenant, not to use the military, by Veto and the USA declined to deliver soldiers charged with crimes against humanity to international justice. Vladimir Putin criticized that the USA used military power to change a government. Many Russians saw the European Union to be in crisis in the face of post globalism, while in Russia the image of Stalin's dictatorship was reconsidered.

When then in 2014 the Ukraine declined the option to become a member of the Eurasian Economic Union, Russia occupied the Crimea, breaking the Budapest memorandum (Nolte et al., 2014). In February 2022, Russia also broke the UN-Covenant and attacked the Ukraine militarily.

Whether Russia will win this war now is uncertain, as of course the question remains, what winning this war does mean or may be interpreted as such. But whatever the outcome—Russia in geostrategy has moved to the east, toward China—to the Shanghai Organization. In case the tendency against military as instrument of power is resumed, Russia most probably may lose influence, since traditionally it has invested more than average in the military. Possibly though Russia might gain influence in the field of export of gas, oil, and iron ore.

9 Germany

Germany had occupied an important place in the history of "really existing socialism" not only because two Germans invented the concept, but also because the incapability of the GDR (which inherited the center of German industry between Berlin, Chemnitz, and Dresden) to keep up with Western Germany was so very obvious and fences along the border to keep the population from running away as well as the Berlin Wall made that difference palpable. An example was visiting my relatives in Leipzig that was a bureaucratic procedure, including a long control at the border and a presentation at the local GDR-town-office, while visiting the Netherlands was a biking-trip.

What did German reunification change? First of all: I had relatives in the GDR and we loved meeting them without bureaucracy. But then? Since quite a couple of East-German industries were not compatible in the globalizing capitalist markets, the global economic effects were quite limited. Questions, whether Germany might again reach a big-power-status, which also were discussed in the West, were off the point. Germany's population increased, the German capital was moved to the East, almost to the border of Poland, a new kind of national cosmopolitism had been developed, was propagated as victorious and German politicians started to define all aims in terms of humanities, but none of the global restrictions (army only under NATO-command, no Veto, etc.) were altered. But German reunification turned out to be the starting point of NATO and EU expansion to the East (Tausch, 1997).

Anyway, the quarter of a century following the end of the SU offered fine conditions for a medium power. Germany was able to diversify its imports of oil and gas from the Anglo-Saxon oligopoly and import cheap Russian gas, even via a pipeline in the Baltic. Also, it profited from the old and regionally close market Russia. Scientific and personal contacts thrived. Since uses of the German military were limited and German troops were controlled by NATO, German conservatives also were ready to discuss concepts like social defense close to pacificism and anyway did not see much sense in high military expenditures.

Intellectual background was that it had taken a long and fierce intellectual struggle to make it common the knowledge in Germany that many Germans had taken part in the crimes committed during the 2nd World War—the holocaust not only against the Jewish people, but also against other ethnic groups of the SU and especially Russians. For a long time, many had believed that *Opa war kein Nazi*.

Political questioning was started by East-European members of NATO and EU. The USA and Arabian absolute monarchs were interested in selling gas in fluid form on the middle-European market, and the Eastern neighbors of Germany feared that the pipelines connecting (respectively planned to connect) Russia and Germany directly would only serve Russia. First, the Ukraine argued that Germany should pay for the Ukrainian losses on thoroughfares, later especially Poland opposed the project, since it paid for Russian economy and military.

As a matter of fact, many German researchers including the author misjudged Russian political aims and methods. I went on translating Russian public texts and criticizing ideological constructions, in which the West was depicted as degenerating, but did assume that the Russian leadership would use such concepts for propaganda and did not itself follow them. And of course, I hoped that the good economic relationship—gas, oil, and coal traded for cars and machines—could go on, although Poland and others reproached Germany for such politics.

By the Russian military attack against the Ukraine this line of analysis was proved as insufficient and lost credibility. The attack is rendering some losses of German influence probable. We do not know yet, how far academic contacts will be restricted—since *Euler* many German researchers have worked in Russian institutions, or together with some—as I have done.

The economic losses will influence of course much more people than some academics. Diminution of trade with Russia, for many centuries a valuable market, will lower German incomes. In case the Ukraine enters the EU, the payments will be considerable, and now, following Brexit, Germany and the Netherlands are the main net-payers. Also, the ongoing remilitarization of international relations is likely to diminish German influence. Germany had good reasons to work for a diminution of military investments. The Russian attack makes a convincing argument though that military means even in the twenty-first century are potent instruments in politics, and Germany structurally cannot compete very much, but will contribute to the part assigned to it by NATO.

10 Some Conclusions

In global politics most important might be the ongoing remilitarization of international relations and the new Arms-race. It does seem to be obvious that the Russian attack on the Ukraine has pushed global development into the direction of a two-polar system—the new West (under hegemony of the USA) and the old East (under hegemony of China).

In global history both within Germany and Russia important contributions to the transitions of our time have been elaborated, from the second industrial to the socialist revolution, from the *Wunderkinder* in physics (keeping in mind of course that these *Wunderkinder* connected to Albert Einstein had to flee to the USA to save their lives) to those in space-traffic, from Goethe to Kafka and Pushkin to Tolstoj.

Mass crimes in German and Russian history must be evaluated in the contexts of general western colonialism and racism (Mann, 2007) but German attempts to murder all people confessing Mosaism and change Russia into a colony as well as Russian attempts to murder the social group enemies of socialism and risk the starvation of millions in order to control that social group of peasants appear to belong to a specific category. The NS-officials were not able to define race concretely and in fact had to investigate the Church-archives for the Arian, that means Christian Grandma to decide who was a Jew. In the campaigns in Eastern Europe religious rituals like circumcision were used as indicators, which enabled a few Jewish POWs to survive as Muslims.

Russia's attempt in 2022 to keep up a lost Empire by force in a period, when in politics Empires had been proved incapable to solve the problems of the twenty-first century and in habitus force had been proved as unbearably destructive in a world densely settled and (more or less) cultivated, sadly enough is in the latter context of politics.

References

Darwin, J. (2008). *After Tamerlane. The rise and fall of global empires*. Penguin.

Donecker, S. (2008). Konfessionalisierung und religiöse Begegnung im Ostseeraum. In A. Komlosy, H.-H. Nolte, & I. Sooman (Eds.), *Ostsee 700 – 2000* (pp. 91–109). Promedia.

Fitzpatrick, S. (1999). *Everyday Stalinism*. Oxford University Press.

Herzen, A. (1969). *Vom anderen Ufer*. Rogner & Bernhard.

Hitler, A. (1935). *Mein Kampf*. Zentralverlag der NSDAP.

Hoffmann, P. T. (2015). *Why did Europe conquer the world?* Princeton University Press.

Kahan, A. (1989). *Russian economic history*. University of Chicago Press.

Kantorowicz, E. (1987). *Kaiser Friedrich der Zweite*. Georg Bondi Verlag.

Kay, A. (2021). *Empire of destruction*. Princeton University Press.

Kennedy, P. (1989). *The rise and fall of the great powers*. Vintage.

Leuschner, J. (1985). *Deutsche Geschichte*. Vandenhoeck & Ruprecht.

Mann, M. (2007). *Die dunkle Seite der Demokratie*. Hamburg Press.

Modzelewski, K. (1969). *Monopolsozialismus*. Hoffman & Campe.

Nolte, H. H. (2005). *Weltgeschichte. Imperien, Religionen und Systeme*. Böhlau.

Nolte, H.-H. (2012). *Geschichte Russlands*. Reclam.

Nolte, H.-H. (2015). *Religions in world- and global history*. Lang.

Nolte, H.-H., Bonwetsch, B., & Schalhorn, B. (Eds.). (2014). *Quellen zur Geschichte Russlands* (pp. 523–539). Reclam.

Nolte, H. H. (2016). The Tsar gave the order and the boyars assented. *The Medieval History Journal, 19*(2), 229–252.

Nolte, H.-H. (2017). *Kurze Geschichte der Imperien* (pp. 394–417). Böhlau.

Nolte, H.-H. (2018). Russland und Deutschland im gegenwärtigen System der Mächte. In H.-H. Nolte & R. Wernstedt (Eds.), *Russlandbilder – Deutschlandbilder* (pp. 108–140). Muster-Schmidt.

Sakwa, R. (2010). *The crisis of Russian democracy*. Cambridge University Press.

Schenk, B. (2004). *Aleksandr Nevskij*. Boehlau.

Schlögel, K. (2017). *Das sowjetische Jahrhundert. Archäologie einer untergegangenen*. Welt. Beck.

Schmidt, C. (1997). *Leibeigenschaft im Ostseeraum*. Boehlau.

Smolarz, E., & Nolte, H.-H. (2023). Slavery and serfdom in Muscovy and the Russian Empire. In D. Pargas (Ed.), *Palgrave handbook on global slavery* (pp. 145–162). Palgrave.

Stökl, G. (1967). *Osteuropa und die Deutschen*. Stalling.

Tausch, A. (1991). *Russlands Tretmühle*. Eberhard.

Tausch, A. (1997). *Schwierige Heimkehr*. Eberhard Verlag.

Varese, E. (2001). *The Russian Mafia*. Oxford University Press.

von Werth, Ch. (2007). Die Ukraine. Vielfalt der historischen Kulturlandschaften als chance. In H.-H. Nolte (Ed.), *Transformationen in Osteuropa und Zentralasien* (pp. 23–40, 126–137). Wochenschau.

Wernstedt, R. (2018). Bilder von Russland, Bilder von Deutschland. In H.-H. Nolte & R. Wernstedt (Eds.), *Russlandbilder* (pp. 39–56). Deutschlandbilder Muster-Schmidt.

Part III
Global Values

Globalization, Islamism, and the Persistence of Religion

Russell A. Berman

Abstract Processes of globalization have included both economic and cultural dimensions, whereby the latter involve secularization and universalism as parts of the long-term program of modernization. Yet in recent decades, countervailing pressures of deglobalization have corresponded to a reassertion of anti-universalist national allegiances as well as a renewal of religious affiliations and identities. Tausch's work shows that an overemphasis on Islamist extremism is as flawed as is ideological decoupling of Islam from Islamism. This topic has been particularly prominent in France, where the responses to Islamism have included both a reassertion of the secularist program of *laicité* and efforts to recover a national Catholicism.

Contradictory forces of globalization and deglobalization are shaping our era. On the one hand, the development of border-crossing technologies, the proliferation of new communications media, and the facilitation of transportation have accelerated processes of internationalization. It has long been a truism that the world has grown smaller, but it continues to shrink. Goods, ideas, services, and people move further and more quickly than in the past. Moreover, this process has been compounded by the growing recognition of challenges that surpass national limits, such as pandemics and climate change, proving the necessity of international collaboration or even governance.

Yet, on the other hand, we face simultaneous evidence of the durability of claims of national sovereignty, most dramatically with the decision of the United Kingdom to leave the European Union. One can cite the vigorous Ukrainian effort as well to preserve national independence against Russian imperialism. Phrased in general terms, we are witnessing the reassertion of post-universal cultural identities, the valorization of forms of localism and retreats from the structures of globalization. In this vein, the breakdown of extended supply chains during the COVID-19 crisis has

R. A. Berman (✉)
Stanford University, Stanford, CA, USA
e-mail: berman@stanford.edu

G. Segell (ed.), *Development, Globalization, Global Values, and Security*,
https://doi.org/10.1007/978-3-031-24513-8_10

produced calls for the relocalization of manufacturing in the interest of the security exigencies of nation states. Programs for internationalism may remain hegemonic, but the appeals to nationhood, nationalism, and subnational regionalism have grown more vocal. Globalization has arguably reached its limits, and a restructuring appears to be under way.

As with all efforts at historical periodization, specifying a precise start date for our current era is an impossible task. Globalization has a long history, but there can be no doubt that the fall of the Berlin Wall on November 9, 1989, marked a caesura, in Europe and beyond, regarding the opening of borders and the speeding up of globalization. The two-bloc world came to an end, especially as Communist China seemed to grow integrated into the international system, despite the tenacity of its party dictatorship. In the wake of the collapse of the Soviet Union, a giddy decade of liberalization followed in the West, and with it the uniquely Western illusion of an end of history. That reality denial itself came to an end on September 11, 2001, when history reasserted itself brutally.

The political and economic liberalism of the 1990s gave expression to a late stage of the optimism of modernism and the idea of a progressive spread of a global culture of enlightenment rationalism and secularization. Even though the attack on the Twin Towers and the Pentagon showed that religio-cultural allegiances still mattered, the responses on the part of the international community, including the War on Terror, continued to carry the banner of globalization. By a decade and a half later, however, with the 2016 Brexit vote and the subsequent election of Donald Trump in November in the same year, the optimism of globalization began to fray. The response to the pandemic in 2020 similarly depended foremost on the sovereign decisions of national governments, rather than on international structures, further indicating the limits of globalization.

Facing that major public health emergency, national leadership acted expeditiously, and often with the telling post-global symbolism of border closures. In contrast, international organizations proved to be slow and cumbersome, often constrained by the political self-interest of powerful member states, as with the influence of China on the World Health Organization. More recently the response to the Russian invasion of Ukraine that began on February 24, 2022 has included extensive economic sanctions that have again disrupted structures of international finance and trade, in effect returning the world to the bifurcated political economy that prevailed before 1989. The global economy is splintering into distinct groups: the sanctioning states of the West, the targets of sanctions, that is Russia, but also Iran, Venezuela, and North Korea, and the many states in between, trying to play the two blocks off against each other, reminiscent of the classic patterns of the Cold War.

This era of globalization and deglobalization between 1989 and 2024 not only involved restructuring the space of political economy but also the frameworks of culture. The enlightenment paradigm of universalization that once defined modernity may not yet have been fully surpassed, but it faces increasingly vigorous competition from claims of a post-modernity, according to which local or regional value systems assert priority over the alleged imperiousness of universalism.

Consequently, distinct national identity can now act as a compelling framework for political solidarity, just as religion may again serve as a source of both personal meaning and participation in communities. The modernist predictions of a disappearance of religion have been proven wrong. The reemergence of national cultural identity communities—as opposed to an international universalism—sometimes takes the form of, for lack of a better term, populist discourses. Populism can have either left or right colorations, just as it can blur the difference between those two traditional ends of the political spectrum. In any case, the erstwhile aspiration for the universal human now confronts the challenge of identity politics, whether it is articulated in terms of race or gender, or nationality or religion.

The symbolic, if not chronological center of this period was, 9/11, with its reassertion of a cultural religious claim against secularism and capitalism. However, Islam is by no means the only example of the contemporary religious cultural resurgence: one can point as well to the Hindutva of Narendra Modi's India, the cultural politics of Catholicism, especially but not exclusively associated with the late Pope Benedict, or the vitality of Christian evangelism in parts of Africa and Latin America. It is plausible that political theological claims associated with Russian Orthodoxy may be playing a role in the ideological framework of Putinism, but at this point, during the war, it is too early to gauge the specific significance of religion in current Russian identity. Whatever the particularities of local formations, this widespread reassertion of religious claims can be described in terms of social theory as deroutinization, from a Weberian perspective or, borrowing from Walter Benjamin, as reauraticization. At stake ultimately is a profound challenge to the inevitability claims associated with secular modernization.

In the shadow of the War on Terror, this discussion of religion and politics has played out especially with reference to Islam. Unfortunately, the debate has often been dominated by exclusive extremes, based on caricatures rather than on empirical data: the claim that Islam is endemically bellicose, and that terrorist violence is therefore the genuine expression of Islamic teaching or, alternatively, the assertion that Islam is fundamentally pacifistic, a religion solely of peace, and that the perpetrators of terror have nothing to do with it, even when they invoke it to justify their deeds.

In contrast, Tausch's contributions have been marked by a persistent commitment to bringing objective, social scientific methods to bear on quantitative data in order to reach tenable conclusions. Regarding the fraught question of the relationship of Islam to violence, he has convincingly demonstrated the validity of a moderate description, rejecting both the Islamophobic insinuation that Islam leads necessarily to terror and the apologetic account that the terrorist violence is incompatible with Islam. Drawing on public opinion polls, he together with Hussein Solomon has shown how very large and sometimes majoritarian segments of Muslim populations criticize the violence of the Islamic State, while nonetheless large minorities, and—depending on the question—sometimes majorities have endorsed violence, albeit not the violence of ISIS, as examined by Solomon and Tausch (2020). Thanks to their work we know that the simplistic assertions at either end of the spectrum are wrong.

Yet equally important as the quantitative light that Tausch shines on this polarizing topic is his cultural evaluation of the fundamental incongruence of terrorist brutality with the core teachings of Islam. This goes far beyond public opinion data and instead addresses the substance of the religio-cultural claims. Of course, he understands that the terrorists see themselves as acting in the name of Islam, but he nonetheless insists on the incompatibility of their brutality with the teachings. In his words, "We should make it clear here that the Islamist terror ideology and practice are against everything which Muslim humanism stands for, namely a successful and just economic system, good governance, respect for human rights, freedom of choice and protection under a specified code of law, and a code by which to manage external relations (with non-Muslims)" (2021, 187–218). To be sure, not all Muslims necessarily endorse all these desiderata, and in particular violent extremists have no interest in implementing a human rights program. Nonetheless, Tausch emphasizes repeatedly that it is wrong to equate the Islamists with Islam or to accept the notion that the most extremist positions are therefore the most Muslim.

On the contrary, he reminds us that the opponents of the extremists are Muslims as well: "To equate 'Muslims' with terrorism is unjust—just recall the heroic example of the Jordanian Air Force pilot Muath Safi Yousef Al-Kasasbeh, who, on January 3, 2015, was burnt alive by ISIS after his F-16 crashed during an operation across ISIS territory. Tausch (2021, 289–34) points out that he, too, was a believing Muslim and a Jordanian patriot."

In the face of hostile representations of Islam, Tausch's insistence on differentiation is vital. In addition, however the argument that Islamic teaching can lead to positive socio-political goals or the gripping formula of "a believing Muslim and a Jordanian patriot" raises the question of the status of religion in any political community. Here the work of Ruud Koopmans provides a useful entry way, insofar as building, like Tausch, on quantitative data, he enquires into the relative dearth of democracies in Muslim-majority countries. Neither Tausch nor Koopmans attempts to make an argument based on theological tenets or holy texts: they do not offer "essentialist" arguments about Islam. They differ however in that Tausch works primarily with public opinion data from the Muslim world, while Koopmans looks at political structures and their evaluation in terms of the Freedom House Index.

Koopmans shows that the wave of democratization that swept the world since the 1970s and especially after the collapse of the Soviet bloc largely bypassed the Muslim world. About one quarter of independent countries in the world have Muslim majorities, and some 25 of them are dictatorships, in contrast to only 22 dictatorships in the remaining three quarters of the world. Nor can one counter that democratization in the Muslim world has merely been slow. On the contrary, processes are moving in the opposite direction. "While in other parts of the world popular uprisings and revolutions in the last fifty years against authoritarian regimes have often led to democratization, revolutions in the Muslim world have almost always resulted in new forms of authoritarianism, either because Islamic fundamentalists seized power by force themselves, as in Iran and Afghanistan, or because the army staged a coup against the threat of a fundamentalist takeover, as in Algeria and Egypt, or because democratically elected Islamists, as in Turkey, have gradually abolished democracy" (2021). He also notes that "one can have a long debate about whether Islam and democracy can theoretically go together, but in practice this marriage has rarely worked out."

Exploring various explanations for this democracy deficit in the Muslim world, Koopmans concludes with the paradoxical observation that there is considerable support for democratic systems, but it is often linked to an insistence that religion play a central role in governance. This may mean that civil law should formally incorporate religious law and that sharia therefore become the basis for jurisprudence. Alternatively, it can imply that non-elected religious leaders should be able to play significant roles in political decision-making. In either case, in the terminology of Mark Tessler, "secular democrats" in the Muslim world are often subordinate to "Islamic democrats," whose policies ultimately place a substantive limit on the possibilities of a genuinely liberal democratization.

Koopmans concludes that "this combination of widespread support for the formal principle of democracy with the simultaneous preference of majorities for a strong role of religion in politics explains a recurring phenomenon in Muslim countries: that elections, even if they are by and large free, have often brought to power Islamist parties and regimes whose aims, and actions run counter to the liberal foundations of democracy. This is what happened after the revolution in Iran in 1979, in the first free elections in Algeria in 1991, in the Palestinian territories in 2006 when Hamas came to power, in the wake of the Arab Spring in Egypt and Tunisia, and in a more gradual way in Turkey under Recep Tayyip Erdogan."

From this perspective, it is the preponderance of religious allegiances that results in the democracy deficit, preventing developments toward the secularism and, potentially, generating explicitly authoritarian regime structures. Yet even if one concedes that the expectation that religion play a role in public life limits liberalization, one has in fact only displaced the analytic question from the observation of insufficient democratization to the need to explain the tenacious grip of religion: is the insistence on the importance of religion merely an expression of backwardness or is it possible that religion in fact provides some substantively positive experience that might otherwise be denied by modernity? In this latter case, the contemporary western encounter with Islam should not be addressed solely as an engagement with the alterity of a different culture or in terms of its putatively violent character but, more importantly, in light of a problematic at the heart of western modernity itself, that is the status of any religion in relation to political form. As liberal modernization, associated with globalization, reaches an impasse, the sacred becomes thematic: the return of the repressed.

The competition of the secular and the sacred has been playing out nowhere more dramatically than in France: once known as the "oldest daughter of the church," it became the homeland of the Revolution, the birthplace of *laicité*, and the incubator of republicanism. Yet as a colonial and then former colonial power, France became the site of extensive immigration, which has in turn elicited a harsh backlash involving a reassertion of nationalism, and not only on the extreme fringe of the political right. It has also witnessed murderous Islamist terrorism, some of which has born an explicitly antisemitic character, reawakening the nightmares of the European twentieth century. How could this atavism seize center stage in France, the land of ultimate modernity? One finds a trenchant analysis of the French condition as well as the contemporary status of religion in Michel Houellebecq's controversial in 2015 novel *Submission* in which he looked ahead to the France of the 2022 and its presidential election, imaging a victory by an Islamist party and a subsequent transformation of France into an Islamic Republic (Houllebecq, 2016).

As historical prediction, the novel obviously failed; there was no major Islamist political party in the 2022 election, and Emmanuel Macron was handily reelected. Yet the genius of the novel lies less in political prediction or in the description of Islam than in the recognition of a profound cultural crisis. The alienated central protagonist represents a France that has lost its tradition and its faith, condemned now to the anomie of social loneliness. He also embodies the loss of national heritage, staged starkly in a visit to the monastery of Martel, a symbol of national Catholicism, as a monument to a victory over Islamic invaders in 732. The novel provocatively asserts the end of Catholic France that survives only as nostalgia: where fully secularized France, according to the narrative, can no longer offer sufficient existential meaning, it is the magnetism of triumphant Islam that may provide a plausible cultural alternative to an empty and obsolete modernity.

Houellebecq's novel is, of course, only a fiction, but it had significant public resonance. In this context it is worth treating its proposal as a credible competitor to Koopman's analysis of Islam and politics. Koopmans measures the Muslim world against western liberal modernity in the sense that its absence of democracy is viewed as a deficit, which he explains by the persistence of religion. In contrast to that judgment on the backwardness of political form in the Muslim world, Houllebecq directs our attention to the contemporary post-religious West, too feeble to generate structures of social cohesion or paradigms of life orientation that could repair a fundamental alienation. In fact, the fictional election described in the novel precisely follows Koopmans' paradigm of a formally free voting that still brings an authoritarian religious party to power. The key difference between the two theses involves the alternative judgments on religion: as an impediment to democracy for Koopmans or as a potential source for life meaning, according to (a perhaps ironic) Houllebecq.

The difference between these two views involves the status of religion, or more broadly, the validity of any positive cultural tradition that surpasses abstract liberal criteria. The logic of modernization has involved the assertion of abstract norms, typically in tension with local loyalties, national patriotism, or denominational specificity. The question that we face at this moment of deglobalization involves the credibility of that liberalism in the face of alternative narratives and value structures. The issue was addressed particularly acutely in Macron's description of French nationality in his speech delivered at the Panthéon of September 4, 2020 (Macron, 2020). France had faced terrorism as well as what became known as separatism, the tendency of parts of the Muslim community to withdraw into distinct values spaces.

Macron would come to oppose this fragmentation of social cohesion by insisting on the priority of shared political identity, as he defined it in his speech which was addressed to a group of diverse immigrants who had just taken on French citizenship: "To be French means first loving liberty passionately. De Gaulle evoked the two-hundred-year-old pact between France and the liberty of the world. Starting today, you are bound by that pact. In our Republic, liberty is a block. It is a liberty to participate in the choice of its leaders and therefore the right to vote, which is however inextricably tied to accepting the electoral decisions, the collective liberty of the people. It is the liberty of conscience, and in particular the secularism (*laicité*), this arrangement unique in the world that guarantees the freedom to believe or not to believe, and which is not separable from a liberty of expression that goes as far as the right to engage in blasphemy."

Macon's invocation of a right to blasphemy is a reference to the 2015 murders of the editorial staff of the satirical magazine *Charlie Hebdo*, provoked by their having published caricatures of the prophet Mohammed. Freedom of speech, according to Macron, trumps any obligation to respect others' beliefs.

Of particular interest in addition however is how Macron's panegyric of the French Republic involves an invocation of a two-thousand-year history, a national past grounded in the age of the Roman Empire. The chronology may seem questionable to those who would date the Republic from the Revolution of 1789 and the overthrow of the monarchy. Yet Macron evidently anchors the liberty of modern France in a premodern tradition, as if abstract modernity cannot survive on its own, without recourse to deeper and more substantial historical resources. That archaic moment represents a de facto concession to Houellebecq's argument: without deeper roots, including a premodern and pre-republican history, the political self-understanding of liberal France ceases to be sustainable. Macron reaches back to the Roman Empire, while Houllebecq invokes Charles Martel and the Battle of Tours of 732. Phrased somewhat differently: without deep roots, modern civil society withers.

Yet despite the appeal to a deep France, Macron's argument centers on secularization and modernity. His alternative to Muslim separatism is not the Catholicism that Houellebecq's hero sought but rather the loyalty of lay citizens. This formal citizenship is devoid of religious commitment which likely puts it at a disadvantage in the competition with Islam. Is faith stronger than patriotism? Or can there be a patriotism devoid of faith? Questions like these are posed by the novelist Jean-Marie Rouart who claims emphatically that liberal secularism is insufficient against Islamist ambitions: "In the face of a very real Islamist threat that has led to violence [. . .] it is interesting to try to raise the level of the debate. It is necessary to inquire calmly into a question that worries those thinkers least inclined to emotional reactions. In contrast to a Christianity drained of its former ambitions, why is it that Islam has not given up its virulent proselytism and instead appears to pose a threat for the future?" Rouart (2021) paints a picture of a weakened Christianity, disheartened priests and empty Churches, that nonetheless face the "party of secularism," that is the advocates of liberal modernity for whom this religious tradition is merely ignorance. Yet unlike religion, secularism provides no comfort "in those tragedies and misfortunes that all men and women undergo, when hearts bleed and despair abound [. . .]."

Religion is not the hindrance to democracy, as Koopmans describes it, but a source of existential solace and meaning. If Catholicism is eliminated by the secularists, so argues Rouart, it is not religion in general that will end. Instead, a "a great religious inversion" may follow, and Islam will take the place of the Church. "To be sure, this is not a matter of a few years, but perhaps in thirty or forty years, when the demographics of the Muslim world in France will have grown exponentially, and when we will have gradually been trapped in a world without horizon, ultra-materialistic, shaped by the multinationals and nourished by consumption, trivia, and pornography."

Rouart continues, "In addition, as history teaches, periods of extreme liberty bordering on anarchy can often lead to a desire for submission. For these reasons, Islam may appear as a moral issue and an opportunity for a spiritual life for all those disappointed with Christianity or no longer convinced by it, even without themselves adhering to secularist tenets. That will be a different revolution, a complete restructuring of our values and our institutions." With this prediction, Rouart predicts a possible realization of Houellebecq's narrative. Yet the conclusion to draw at this point is not the inevitability of a Muslim wave across Europe, a great replacement. The point rather is that religions and traditions can be carriers of life validation and profound meaning that are vital to the thriving of society. Banishing religion from public life can only enervate culture and weaken identity.

As President, Macron has advocated a two-track program to confront Islamism: first, a prohibition of the separatism of parallel societies that claim to stand outside of French jurisdiction, and second and more generally, the insistence on the secularization of public space. If the former involves upholding the principle of equality before the law, the second entails the logic of banning religion into the private sphere and thereby diminishing it. Rouart's response treats this *laicité* as a retreat from French traditional national identity and, furthermore, as a losing proposition: France cannot resist the magnetism of Islam by concealing its own Christian or Judeo-Christian heritage. In the wake of Macron's victory in the 2022 presidential election, his second tenure in office will provide an opportunity to test this claim. Can Macron's universalist France provide a compelling bulwark against Islamism? If not, illiberal programs of particularist identity politics, of the sort advocated by defeated candidates Marine Le Pen and Eric Zemmour, will continue to gain credibility in political debate in this era of deglobalization.

References

Houllebecq, M. (2016). *Submission*. Picador.

Koopmans, R. (2021). *Why has democratization bypassed the Muslim world? The Caravan Notebook*. Retrieved August 1, 2022, from https://www.hoover.org/research/why-has-democratization-bypassed-muslim-world

Macron, E. (2020). The speech of the President of the Republic on the Occasion of the 150th Anniversary of the Proclamation of the Republic, in the Panthéon. *TelosScope.* Retrieved August 1, 2022, http://www.telospress.com/macron-on-french-nationality-the-pantheon-speech/

Rouart, J.M. (2021) Secularism is an illusory defense against the Islamist will to conquer. *TelosScope.* Retrieved August 1, 2022 http://www.telospress.com/secularism-is-an-illusory-defense-against-the-islamist-will-to-conquer/

Solomon, H., & Tausch, A. (2020). *Islamism, crisis and democratization: Implications of the world values survey for the Muslim world.* Springer.

Tausch, A. (2021). Social attitudes fuelling Islamist terrorism. In H. Solomon (Ed.), *Directions in international terrrorism: Theories, trends and trajectories* (pp. 17–34). Palgrave Macmillan.

Studies on Global Catholicism

Stanisław Obirek

Abstract This essay starts by highlighting the author's collaboration with Arno Tausch. An analysis of decades of data collected by Ronald Inglehart's World Values Survey, a global social science research initiative in the 1970s, on the value system, provided hard data on the condition of Catholicism. The results allow for the formulation of a research hypothesis on the almost global erosion of Catholicism. This word almost is very important as it highlights attempts to reverse the decomposition of institutions through healing initiatives. It should be noted right away that they apply not only to Catholicism, but also to other denominations and religions. So Roman Catholicism is not the only institution dealing with a crisis that has been going on for decades. One can even say that it applies to all institutional religions, including other denominations of Christianity or the other monotheistic religions as Islam or Judaism.

1 Almost Global Erosion of Catholicism

When Arno Tausch invited me a few years ago to contribute to the book Global Catholicism, Tolerance and Open Society: An Empirical Study of the Value Systems of Roman Catholics (Tausch & Obirek, 2020), I did not realize then that this would be the beginning of my reflection on the global condition of contemporary Catholicism. Moreover, an analysis of decades of data collected by Ronald Inglehart's World Values Survey, a global social science research initiative in the 1970s, on the value system, provided hard data on the condition of Catholicism.

The title of our book talks about tolerance and an open society, which we really dealt with, but we also devoted a lot of attention to anti-Semitism, xenophobia, and homophobia, which are not only present in the mentality and attitudes of many Catholics, but even we have to do with the intensification of these behaviors in some

S. Obirek (✉)
American Studies Center, University of Warsaw, Warsaw, Poland
e-mail: s.obirek@uw.edu.pl

G. Segell (ed.), *Development, Globalization, Global Values, and Security*,
https://doi.org/10.1007/978-3-031-24513-8_11

countries. These data only confirm the research intuitions that have been present in numerous books by sociologists, anthropologists, philosophers, and even theologians for years.

The results allow for the formulation of a research hypothesis on the almost global erosion of Catholicism. This word almost is very important as it highlights attempts to reverse the decomposition of institutions through healing initiatives. They are undertaken by individuals, who often must deal with the hostile reaction of church institutions. Some of these people stay in the Church trying to change it from within (Dupuis, Gutierrez, Haight, Martin, Phan), many people leave the Church changing their religion (Curran, Smith-Leukel, Jordan), others perform an act of apostasy (Ehrman, Polak) or they will stand in opposition to the official enunciations and gestures of the representatives of the Church (Martel, Nowak, Obirek). These are individual initiatives and do not fundamentally change church structures.

Hence, in my opinion, the main direction of changes is negative for Catholicism, which is confirmed by almost universal secularization, as recently written by Ronald Inglehart (2021). It is the author whose numerous publications allow us to better understand the global changes in religiosity, in which the decline of the authority of institutional religions plays a significant role. In his opinion, there has been an acceleration of secularization processes in recent years. While most countries became more religious from 1981 to 2005, from 2007 to 2020 the vast majority became less religious. One indicator of this accelerated secularization is the emergence in recent years of a significant none group in a country traditionally considered immune to secularization processes like the USA (Burge, 2022).

It should be noted right away that they apply not only to Catholicism, but also to other denominations and religions. So Roman Catholicism is not the only institution dealing with a crisis that has been going on for decades. One can even say that it applies to all institutional religions, including other denominations of Christianity or the other monotheistic religions as Islam or Judaism. Today's crisis also affects the religions of the Far East, such as Buddhism, Hinduism, and Shintoism. In my reflections, however, I will limit myself to Catholicism, in which, however, it is possible to recognize elements found in other religions as well.

2 Church Control System

The Protestant sociologist of religion, Peter L. Berger, in his book The Heretical Imperative: Contemporary Possibilities of Religious Affirmation, acutely noticed at the end of the 1970s that the intellectual curiosity of Christians, which was suppressed for centuries, had gained a new status: it had become the requirement of the moment. Berger refers to the etymology of the word heresy, which means the ability to choose and decide according to the choice made. So today, not only is heresy not to be avoided, but heretical thinking has also become the requirement of the moment (Berger, 1979). Berger's intuition runs counter to the Catholic understanding of theology, which is based on obedience to church authority. At least that was how it was understood until the Second Vatican Council (1962–1965), which, in such declarations as Dignitatis Humanae and Nostra Aetate, restored the primacy of freedom of conscience and the right of a Catholic to decide his or her own world view, including the religious one.

However, despite these revolutionary changes, it should be noted that they remained a dead letter in ecclesiastical practice that continued to appeal to the argument of force and hierarchy, and especially the Roman Curia, continued to discipline theologians. In this context, it became understandable the question asked by one of the dissident theologians, Hans Küng: whether the Church can still be saved. Küng referred not only to his own experiences (as we know, in 1979 he was deprived of the right to teach Catholic theology by the Congregation for the Doctrine of the Faith), but also invoked the entire control system paralyzing the freedom of theological research in the Catholic Church. Küng gave a positive answer to the title of his book, because in his opinion it is possible to develop solutions within the Catholic tradition that would respect the intellectual and moral autonomy of the Catholic believer (Küng, 2011).

However, in the post-conciliar period, reforms were stopped already during the pontificate of Paul VI, especially through the encyclical Humanae Vitae announced by this pope in 1968, which closed the discussion on human sexuality and women's reproductive rights. As Charles Curran, deprived of the right to teach Catholic theology like Hans Küng, wrote, without changing the doctrine expressed in Humanae Vitae, the Church risks further losing its credibility and will not be able to honestly face the other challenges of the present time (Curran, 2018).

Mark D. Jordan went even further in his criticism of the church system. Jordan wrote many times about the oppressive nature of Catholic theology on homosexuality and about the Catholic Church as a homophobic institution with a clear homoerotic overtone. In his opinion, it is not about revising or changing church doctrine, but about transforming power structures (Jordan, 2014, 2). For Jordan, the presence of this paradox (theoretical homophobia and homoeroticism practiced by the clergy) was the reason for his departure from the Church. However, knowledge of the history of this paradox allows him to write more books deconstructing the hypocrisy of Catholicism (Jordan, 2002).

Mark Jordan's intuition is confirmed by a richly documented book by the French sociologist Frederic Martel, who, using the examples from hundreds of interviews with representatives of the clergy (cardinals, bishops, priests, and seminarians) and former priests, showed how deeply homosexual culture is rooted in this environment (Martel, 2019). And the American historian Dyan Elliott, in an extremely meticulous analysis of historical documents, showed that the problem of homosexuality (and pedophilia) in the Church has deep historical roots (Elliott, 2020).

In this context, one can mention the dramatic changes in the Irish Church, which experienced a real collapse under the influence of pedophile scandals involving the clergy. The former priest Tony O'Dwyer described them in detail, presenting the results of several reports devoted to this matter (O'Dwyer, 2019). A specific complement to the problems related to the clergy is the still unresolved issue of the current celibacy, which generates several problems. These issues are perfectly presented in the book by Italian sociologist Marco Marzano (Marzano, 2021).

Many Catholic theologians, despite constant criticism from the Congregation for the Doctrine of the Faith (CDF), which is the direct heir of the infamous inquisition, decided to stay in the Church. Such is the case of Gustavo Gutierrez, the father of the theology of liberation, or Jacques Dupuis, one of the most interesting theologians of interreligious dialogue, and the author of a new understanding of Christology by Roger Haidt. Again, there is a paradox in this context. Well, thinkers who develop innovative interpretations of the Catholic tradition within the church structures are treated with suspicion by the Vatican control system, and key positions (bishops, cardinals) are filled not only by people faithful to the system, but often persecuting original theologians.

The theology of liberation, initiated by the book by Gustavo Gutierrez (1988), met with strong criticism from the CDF, which has repeatedly taken a critical position toward this theology. For the last time in 2006, he criticized the books of the Jesuit Jon Sobrino, one of the most distinguished representatives of theology of liberation, only because he proposed a new understanding of Jesus as a liberator not only in theological but also in a social sense (Sobrino, 1993).

An important complement to this intra-Catholic theological conflict is the synthetic approach to the problem of liberation theology in the context of global capitalism, developed by the interdisciplinary team of Andreas Müller, Arno Tausch, Paul Michael Zulehner, and Henry Wickens. This book shows, on the one hand, how important a phenomenon theology of liberation was and is on a global scale, and on the other, it points to the shortsightedness of the Vatican's politics (Müller et al., 2000).

The situation of the Belgian Jesuit Jacques Dupuis, who worked almost all his life as a missionary in India, was particularly dramatic. In the 1980s he returned to Europe and lectured at the Pontifical Gregorian University. It was the proximity of the Vatican that drew the attention of the officials of the CDF, which accused him of heresy. Moreover, the head of this Congregation, Cardinal Joseph Ratzinger, in 2000 announced the declaration Dominus Iesus, directed directly against the new interpretation of the saving action of Jesus as proposed by Dupuis. A positive effect of this censorship activity was the new book by Dupuis, in which he once again laid out his concept, pointing out how much he was in the centuries-old tradition of the Church (Dupuis, 2002).

The American Jesuit Roger Haight faced similar harassment, who in turn proposed to understand Jesus as a symbol of God, which also caused anxiety among Vatican censors (Haight, 2002). The names cited are only a few examples of theologians subjected to real persecution in the post-conciliar period. This was especially true of the time related to the pontificates of John Paul II and Benedict XVI. The former was rightly called by Karl Rahner the *Winterliche Zeit*, and one of his biographers wrote about the dark face of the papacy (Cornwell, 2004). The controversy surrounding Benedict XVI is far from over, and his role in covering up the pedophilia of the clergy is yet to be clarified.

Thus, John Paul II, together with his closest advisor, Joseph Ratzinger, almost from the very first moment of his pontificate, that is from 1978, focused primarily on combating new trends in theology and pushing forward his conservative vision of man, especially human sexuality. Not everything can be documented, although there are many documents. But much more important is the creation by these two prominent hierarchs of the entire system of informing, mutual distrust and flattery, which successfully paralyzed the creative ferment in the Catholic Church that began at the Second Vatican Council in the mid-1960s.

It seemed that thanks to the revolutionary decisions of John XXIII, who returned to favor previously persecuted theologians such as Henri de Lubac, Yves Congar, Jean Danielou, and others, and opened the windows and doors of the Church to speak of contemporary culture. It was then that openness and dialogue became the most important concepts in Catholicism. Unfortunately, when Karol Wojtyła assumed the papal office, the times of the anti-modernist crusade that lasted over a hundred years from the mid-nineteenth century to the sixties of the twentieth century returned. Only in this context can one understand the irreparable damage suffered by local Churches, especially in South America. A place of renewal and conciliar reform was followed by a period of regression and a return to confrontation with the world.

In this context, one can speak of the continuing legacy of the Inquisition and the Holy Office, which over the centuries destroyed independent thinking in the Catholic Church. It did not end, as noted by the eminent historian Edward Peters with the successes of the Reformation and did not become a myth (Peters, 1988). So, it still effectively paralyzes independent thinking in this institution. On the other hand, it must also be said that despite the continuing system of control in the Catholic Church, there are effective attempts at changes that are worth remembering.

3 Slots in the System

The American Jesuit and historian John O'Malley in his monograph on the Second Vatican Council wrote that it was primarily a language event. O'Malley says that what really happened was a change of language. In fact, in the language there has been a true "turn" of the Catholic Church. It is not only about introducing new words into the dictionary of Catholic theology, such as freedom of conscience, dialogue, human dignity, cooperation with other faiths, respect for other religions or recognizing the existence of lay people in the Church. The language indicated and induced a change in values or priorities. To this extent it indicated and induced internal conversion, "which means the radical change that takes place under the influence of profound experience. It is usually associated with a change in the current lifestyle and departure from previous practices".

In extreme cases, there is a break with the past. The French Jesuit Michel de Certeau, a keen observer of the changes taking place in the Church in the 1960s and 1970s and an active participant in them, spoke of a renewal break (*rupture instauratrice*). Thus, we are dealing here with the classic situation of a paradigm shift under the influence of falsifying new scientific data, about which Thomas Kuhn wrote years ago (Kuhn, 1970). This change is a shift from the model of exclusive theology to that of inclusive, or even pluralist theology.

For the Catholic theologian, the change made at the Second Vatican Council is binding, and in this new context he or she is aware that it is impossible to absolutize his or her own point of view but is open to changing and correcting his or her own position under the influence of new data. According to the German theologian Perry Schmidt-Leukel, one can even speak of the fractal structure of theology. In other words, the realization of the multiplicity of religions and the inclusion of this new awareness in theological reflection mean that in other Christian denominations and other religions we see a necessary complement on the way to salvation (Schmidt-Leukel, 2017).

What was this turn in the case of Roman Catholicism? The consequences are more serious than you might think. Acquiring a new language so thoroughly as to live in harmony with other religions must entail an internal transformation. It is about much more than learning new terms for your old concepts. Correctly using a new language means fully embracing the values and sensibilities of a culture other than your own. So, the Church, in the light of the above statement, should also become different. And she became different. Such a change took place in countries where the conciliar documents were not only translated, but also implemented in the daily life of church communities.

It happened in South America, mainly thanks to the so-called basic communities and their leaders—liberation theologians. This is the case in Asia and Africa where Christianity is one of many religions, and Catholics in many countries are in a vast minority. It was for them that opening to other faiths and religions marked the beginning of a new chapter; it is not confrontation but ecumenical and interreligious dialogue that determines their faith. Similarly, Catholics of North American countries, especially the USA and Canada, and in Western European countries, have discovered a partner in dialogue in modern culture (O'Malley, 2008).

It is into this gap created as a result of the revolutionary changes that took place at the council that new theological proposals entered. Since the moment when Jorge Maria Bergoglio became pope in 2013, taking the name of Francis, these changes have the official support of the Vatican. Out of necessity, I can only mention a few of them as harbingers of deeper processes, the consequences of which we are not yet aware of today. One of them is American Jesuit Francis X. Clooney, who deal with interreligious dialogue. Clooney continues the best traditions of the Jesuits dating back to the sixteenth and seventeenth centuries. It was then that the Jesuits developed modern strategies for contact with other cultures and religions, the basis of which was mutual knowledge and respect for a tradition different from my own. However, in the sixteenth and seventeenth centuries, the intuitions of Matteo Ricci and others were rejected by Rome. Only in the twenty-first century is it possible to develop them creatively (Clooney, 2010).

An exceptionally interesting proposal was developed by the former Jesuit Ewert Cousins, who, referring to the classic theory of Karl Jaspers about the Axial Age, proposed an interpretation of civilization changes in the twenty-first century as a new Axial Age, the center of which is Jesus Christ (Cousins, 1998). The new ideas were proposed also in the field of ethics. Working for years at Loyola University Chicago, the German theologian Hille Haker proposed something more than just another version of the textbook approach to Christian ethics. Her critical and political ethics is an attempt to understand why Catholic ethics has found itself in a deep crisis and asks questions about the possibility of overcoming it.

Haker enters a fruitful dialogue with the contemporary humanities by taking up and creatively developing its most interesting achievements, but at the same time pointing to its immanent limitations. The author's most important achievement is the proposal of political ethics that draws not only from the source inspirations of Christianity, but also from the context of postmodernism and post-secularism (Haker, 2020).

The case of the American nun Elizabeth Johnston, who challenged American bishops who criticized her innovative approach to theology, is very interesting. Not only did she not withdraw from her feminist conceptions of God and the Catholic tradition, but she accused them of ignorance and encouraged them to study their own traditions thoroughly (Johnston, 2007). A very significant contribution to the development of Catholic theology is made by the American theologian of Vietnamese roots, Peter C. Phan, who for years has convincingly shown that one can be a Catholic who belongs to many religions at the same time (Phan, 2004).

Also, thanks to the numerous publishing initiatives of Peter Phan, Catholics learn that the opening of Pope Francis to refugees and emigrants belongs to the essence of Catholicism (Theology of Migration in the Abrahamic Religions, 2014). It is also worth mentioning the many initiatives of the Jesuit James Martin, editor of America Magazine, who is an adviser to Pope Francis on LGBT people. It was Martin who wrote a book on building bridges between LGBT communities and Catholicism (Martin, 2018). As you can see, the slots that appeared in the church system at the Second Vatican Council are quite numerous and there are still daredevils who can penetrate them and propose a more human face of Catholicism in the twenty-first century.

4 The Case of Polish Catholicism

In 2018, the independent American research company Pew Research Center published comparative data on the religiosity of people under 40 and over 40 in 108 countries around the world. To the surprise of many observers, Polish youth turned out to be a world leader in the secularization process. Weekly participation in the masses is declared by 55 percent Poles and Polish women over 40 years of age. Among the younger ones—only 26 percent. So, the difference is 29 percentage points. Religion is "very important" to 40 percent Older and only 16 percent Younger. This is a generational decline of 23 percentage points (The Age Gap in Religion Around the World, 2018).

However, by these results were not surprised Polish sociologists of religion, who for years have been carrying out systematic research on the changes in the religiosity of Polish youth. The results of the research by Janusz Mariański, associated with the Catholic University of Lublin, clearly indicate a sharp departure from religious practices by Polish youth (Mariański, 2018). Similarly, Józef Baniak from the University of Poznań points to specific reasons for the secularization processes that Polish youth undergoes. In his opinion, contrary to expectations, it was not young Polish Catholics who influenced the growth of religiosity in secularized Western Europe, but they themselves adopted the behavior patterns of their peers from that part of Europe (Baniak, 2022, 884).

Of course, the gaping secularization of Polish youth is only one of the symptoms of the deep crisis in which Polish Catholicism found itself, for which the most important test was the turning point of political change in 1989. From then on, the authority of the Church began to decline systematically. This is even more surprising that at the end of communism, and especially since 1978, when Karol Wojtyła was elected pope and took the name of John Paul II, the Church enjoyed enormous social support. Especially due to the support of the Polish pope of opposition movements during his first visit to Poland in 1979. The paradox of Polish Catholicism is that its spectacular success in the first years of John Paul II's pontificate was at the same time the source of its crisis, and even its fall. In his classic book on the presence of religion in public space, Jose Casanova presented an example of Polish Catholicism from the 1980s as a case of a happy combination of religion and social aspirations in the context of political enslavement (Casanova, 1994).

On the threshold of the third decade of the twenty-first century, one can see how the politicization of religion has become one of the main inhibitors of civilization and democratic processes in the former countries of the communist camp, including Poland. Polish bishops, who were not only unable to accept civilization changes, but also systematically rejected them, played a particularly negative role in this process. These processes were presented in the book by Artur Nowak and Stanisław Obirek on the abuse of power in the Church (Nowak & Obirek, 2021). Tomasz Polak presented a systematic criticism of not only the politicization of Catholicism, but also its inability to make any changes, while analyzing the church system (Polak, 2020).

The even destructive function of Catholicism, and not only in the Polish context, was presented by Agnieszka Graff and Elżbieta Korolczuk in a book devoted to specific crusades unleashed by many Catholic activists, bishops, and theologians against gender (Graff & Korolczuk, 2022). These authors' special merit is paying attention to the close ties of Catholicism with right-wing populist movements and their mutual support. However, the harmfulness of combining religious fundamentalism with right-wing populism could not have been so overwhelming were it not for the clear domination of the Catholic Church in the Polish public space.

They were piercingly diagnosed by the Polish philosopher Barbara Stanosz, who wrote many times about Polish democracy in the shadow of the Church (Stanosz, 2004). One of the features of Polish Catholicism that largely explains its inclination toward right-wing parties is its anti-Semitism. It was especially visible in the interwar period, and what is worse, it showed its dangerous face when fascism came to the fore in some Catholic countries in Western Europe (Germany, Spain, Italy, Hungary). Fascism also found numerous supporters among Polish Catholics, as expressed in the Catholic press (Modras, 1994). However, a particularly embarrassing episode in the history of Polish Catholicism of that period (completely repressed from the memory of the Jesuit Order) is organic anti-Semitism, represented by the extremely influential Jesuit Włodzimierz Ledóchowski, who was the general of the order in 1915–1942. It was Ledóchowski, according to historian David Kertzer, who was largely responsible for bringing the Jesuits and the papacy closer to fascism (Kertzer, 2014).

So far, Polish society has a clear problem to free itself from this shadow. However, I would like to end on an optimistic note. Well, the Catholic Church is not only losing influence in Polish society, especially among the younger generation, but also other denominations appear, attracting more and more former Catholics. It cannot be ruled out that Polish society will also experience religious and ethnic pluralism known from the history of the Polish-Lithuanian Commonwealth. The traces of this religious pluralism could be found also in present-day Poland (Pasieka, 2015). Also, the increasingly full integration in the structures of the European Union and the increase in economic security will confirm the validity of Inglehart's hypothesis of the global growth of secularism (Inglehart, 2021).

5 What Does the Future Hold?

Of course, today's Polish Catholicism is an extreme example of the politicization of religion and should not be treated as a valid model, much less as a long-term one. On the contrary, it is rather an exception, and everything indicates that it is undergoing internal erosion, as has already been mentioned. To conclude this brief and necessarily schematic overview of some elements of the transformation processes which global Catholicism is undergoing, I would like to dwell on certain tendencies that seem to define the future of the Roman Catholic Church. The American Vaticanist John Allen, writing about the future of Catholicism in 2011, pointed to ten trends that are changing the Church. Among them were evangelical Catholicism, the growing role of the laity, Islam, globalization, and ecology (Allen, 2009). In his manner of holding the papal office since 2013, Francis tries to pay attention to these trends.

However, in my opinion, a new and much more important challenge to Catholicism has arisen now. They are growing pluralism, post-secularism, an individualistic approach to religion, a positive view of atheism as a kind of mirror in which the believer can look at. The German sociologist Ulrich Beck) wrote about religious individualism, or even about the emergence of tendency to define of God in our own terms (Beck, 2010). Pluralism as an important determinant of the understanding of religion in the twenty-first century was described by Peter Berger in his last book (Berger, 2014). Michele Dillon looked at the pontificate of Francis in the post-secular key demonstrated how promising this pontificate is for the future of the Church (Dillon, 2018). For many researchers, it is post-secularism that is the greatest challenge, but also an opportunity for religion (Obirek, 2019).

Finally, to define God's return after the period of atheism, the Irish philosopher Richard Kearney coined a special term antitheism (Kearney, 2010), which is beginning to gain more and more popularity. However, the phenomenon that arouses the greatest interest of sociologists of religion and theologians is the rapidly growing group of "nones," that is people who do not admit to belonging to any institutional form of religion and, at the same time, do not renounce religion as such. It has already been described in the book by Ryan Burge (Burge, 2021). However, the phenomenon has been noticed already several years ago and for many researchers it indicates a new quality in the approach to both religion and its institutional forms (Obirek, 2016). In this context, the question that in 1935 Joseph Stalin asked the French politician Pierre Laval, who offered him cooperation with the Vatican: "The Pope? How many divisions has he got?" takes on a new meaning. Catholicism, and with it the papacy, has almost completely eroded, so the question arises whether there are slots that will allow it to revive. After all, this is a completely different story that we cannot take up here.

References

Allen, J. L., Jr. (2009). *The Future Church. How ten trends are revolutionizing the Catholic Church*. Doubleday.

Baniak, J. (2022). *Religijność i moralność młodzieży polskiej w latach 1956–2018 na tle przemian cywilizacyjnych i społecznych. Studium socjologiczne*. Wydawnictwo nauk Społecznych i Humanistycznych Uniwersytetu AM w Poznaniu.

Beck, U. (2010). *A god of one's own: Religion's capacity for peace and potential for violence*. Polity Press.

Berger, P. L. (1979). *The heretical imperative. contemporary possibilities of religious affirmation*. Anchor Press.

Berger, P. L. (2014). *The many altars of modernity. Toward a paradigm for religion in a pluralist age*. De Gruyter Press.

Burge, R. (2021). *The nones. Where they came from, who they are, and where they are going*. Fortress Press.

Burge, R. (2022). *How America's youth lost its religion in the 1990s*. Retrieved August 6, 2022, from https://www.ncronline.org/news/opinion/how-americas-youth-lost-its-religion-1990s

Casanova, J. (1994). *Public religions in the modern world*. University of Chicago Press.

Clooney, F. X. (2010). *Comparative theology. Deep learning across religious borders*. Wiley-Blackwell Press.

Cornwell, J. (2004). *The pope in winter. The dark face of John Paul II's papacy*. Viking.

Cousins, E. H. (1998). *Christ of 21st century*. Continuum Books.

Curran, C. (2018). Humanae vitae: Fifty years later. *Theological Studies, 79*(3), 520–545.

Dillon, M. (2018). *Postsecular catholicism: Relevance and renewal*. Oxford University Press.

Dupuis, J. (2002). *Christianity and the religions. From confrontation to dialogue*. Orbis Books.

Elliott, D. (2020). *The corrupters of boys: Sodomy, scandal, and medieval clergy*. University of Pennsylvania Press.

Graff, A., & Korolczuk, E. (2022). *Anti-gender politics in the populist moment*. Routledge.

Gutierrez, G. (1988). *A theology of liberation*. Orbis Books.
Haight, R. (2002). *Jesus symbol of god*. Orbis Books.
Haker, H. (2020). *Towards a critical and political ethics. Catholic ethics and social challenges*. Schwabe Verlag.
Inglehart, R. (2021). *Religion's sudden decline: What's causing it, and what comes next?* Oxford University Press.
Johnston, E. A. (2007). *Quest for the living god*. Bloomsbury.
Jordan, M. D. (2002). *The silence of sodom. Homosexuality in modern catholicism*. University of Chicago Press.
Jordan, M. D. (2014). Talking about homosexuality by the (Church) rules. In J. P. Hornbeck & M. A. Norko (Eds.), *Inquiry, thought, and expression* (pp. 34–51). Fordham University Press.
Kearney, R. (2010). *Anatheism. Returning to god after god*. Columbia University Press.
Kertzer, D. I. (2014). *The Pope and Mussolini. The secret history of pius XI and the rise of fascism in Europe*. Oxford University Press.
Kuhn, T. (1970). *The structure of scientific revolution*. The University of Chicago Press.
Küng, H. (2011). *Ist die Kirche noch zu retten?* Piper.
Mariański, J. (2018). *Kondycja religijna i moralna młodzieży szkół średnich w latach 1988–2005–2017*. Adam Marszałek.
Martel, F. (2019). *In the closet of the Vatican. Power, homosexuality, hypocrisy*. Bloomsbury.
Martin, J. (2018). *Building a Bridge: How the Catholic Church and the LGBT community can enter into a relationship of respect, compassion, and sensitivity*. Harper Collins Publishers.
Marzano, M. (2021). *La casta dei casti. I preti, il sesso, l'amore*. Bompiani.
Modras, R. (1994). *The Catholic Church and Antisemitism, Poland 1933–1939*. Harwood Academic Publisher.
Müller, A., Tausch, A., Zulehner, P. M., & Wickens, H. (2000). *Global capitalism, liberation theology, and the social sciences: An analysis of the contradictions of modernity at the turn of the millennium*. Nova Science Publishers.
Nowak, A., & Obirek, S. (2021). *Gomora. Pieniądze, władza i strach w polskim Kościele*. Agora.
O'Dwyer, T. (2019). *Betrayal by silence. The collapse of the Irish Catholic Church. Has it a future?* Print Bureau.
O'Malley, J. W. (2008). *What happened at Vatican II*. The Belknap Press.
Obirek, S. (2016). Nones – A new religious category, a new form of American Pragmatism? *Discourses on Culture, 5*(2), 145–153.
Obirek, S. (2019). The challenge of post-secularism. *Journal of Nationalism, Memory & Language Politics, 13*(2), 239–250.
Padilla, E., & Phan, P. C. (2014). *Theology of migration in the Abrahamic religions*. Palgrave Macmillan.
Pasieka, A. (2015). Hierarchy and pluralism. Living religious difference in Catholic Poland. .
Peters, E. (1988). *Inquisition*. Macmillan.
Pew. (2018). *The age gap in religion around the world*. Pew Research Center.
Phan, P. C. (2004). *Being religious interreligiously. Asian perspective on interfaith dialogue*. Orbis Books.
Polak, T. (2020). *System kościelny czyli przewagi pana K*. Wydawnictwo nauk Społecznych i Humanistycznych Uniwersytetu AM w Poznaniu.
Schmidt-Leukel, P. (2017). *Religious pluralism and interreligious theology*. Orbis Books.
Sobrino, J. (1993). *Jesus the liberator. A historical theological reading of Jesus of Nazareth*. Orbis Books.
Stanosz, B. (2004). *W cieniu Kościoła, czyli demokracja po polsku*. Instytut Wydawniczy Książka i Prasa.
Tausch, A., & Obirek, S. (2020). *Global Catholicism. Tolerance and open society: An empirical study of the value systems of Roman Catholics*. Springer.

Tausch and Political Controversies in Austrian Media: "Political Islam" in the Austrian Debate

Farid Hafez

Abstract In this essay, the author Farid Hafez reviews a discussion in the daily Die Presse on "political Islam" in Austria, in which he and Arno Tausch were involved. Discussing the main arguments of both camps that were involved, Hafez tries to answer the question, how a reasonable discussion on "political Islam" can take place in a context, where the notion of "political Islam" has been mobilized by the governing New Austrian People's Party led by Sebastian Kurz to serve anti-Muslim racism. By drawing on the example of the Islamic Federations in Austria, which traces its roots back to Necmettin Erbakan, the founding father of political Islam in Turkey, Hafez discusses two different approaches how this religious movement is currently assessed in terms of its origins. The author argues that a critique of racism is ultimately a critique of power structures and since Muslims in powerful positions are part and parcel of these structures, aiming to defend their privileges, a critique of anti-Muslim racism can and even must go hand in hand with critique of these political structures.

1 Introduction

Like other Austrian political scientists such as Anton Pelinka or Peter Filzmaier (Tausch, 2021a), Arno Tausch was active in the media debates as well. In June of 2022, I was invited to contribute to Arno Tausch's Festschrift. This was preceded by initial email correspondence with my fellow Austrian political science colleague Tausch after we had engaged in an argumentative exchange in the Austrian daily newspaper Die Presse a little less than two years earlier (Die Presse, 2020), which summarized some of the findings of Tausch in his previous much more extensive studies (Tausch, 2021c, 2021d).

F. Hafez (✉)
Global Studies Program, Williams College, Williamstown, MA, USA

Bridge Initiative, Georgetown University, Washington D.C., USA
e-mail: fh6@williams.edu

G. Segell (ed.), *Development, Globalization, Global Values, and Security*,
https://doi.org/10.1007/978-3-031-24513-8_12

Tausch, who, as this Festschrift shows, has distinguished himself primarily with academic scholarship discussing quantitative data from the World Values Survey and the Arab Barometer, published a replica of my article in the newspaper. Other than Tausch, I work primarily with qualitative data. However, the debate in the daily newspaper was not limited to our commentary, nor to any commentary written by me. There were, in their entirety, eight comments dealing with the so-called political Islam. As stated in the paper's editorial policy, the daily newspaper Die Presse represents a "bourgeois-liberal outlook on an elevated level" (Die Presse, 2010). It is owned by the Styria Media Group (n.d.), which in turn was founded by the Catholic Church. Along with the daily Der Standard, which sees itself as liberal but far from any conservatism (Der Standard, 2022), it is considered one of the two quality daily newspapers in Austria.

The context of the debate was the first violent attack on the soil of the Republic of Austria, which was on the account of a former Daesh sympathizer and cost four lives on November 2, 2020. A week later, a major raid planned for the night of the attack took place, in which, according to the then Interior Minister, "the roots of political Islam were cut" (Seeh, 2022). This raid was the largest in Austria's history since 1945 but was ruled illegal several months later by the Graz Higher Regional Court (Ibid.). Previously, in July of the same year, the Documentation Center for Political Islam (Bridge Initiative Team, 2021) had been established. It served the purpose of observing and subsequently problematizing this phenomenon, which was identified by the conservative Austrian People's Party (ÖVP) under the leadership of Sebastian Kurz (Bridge Initiative Team, 2020b) as one of the greatest challenges facing society (BKA, 2020). In the following, I would like to give a brief summary of the most important arguments that were used to make the fight against the so-called political Islam appear legit and necessary. In a second step, I will elaborate on the criticism of the government's work. In a third part, in addition to my own argumentation, I will illuminate blind spots in the discussion and elaborate on them.

2 The (Party-Political) Assumption: Political Islam as a Danger

As I have elaborated in a paper, the concept of political Islam has been systematically and strategically introduced in the context of Austrian domestic politics, especially by the ÖVP (Hafez, 2022). While the right-wing FPÖ has been problematizing Islam as a religion since as early as 2005, it is the ÖVP that introduced the term "political Islam." In its election program for the 2017 National Council elections, it makes three claims. Political Islam is a "breeding ground for violence and terrorism [...], a totalitarian system with the ambition to produce parallel legal systems," and most recently promoted from the outside to "undermine our values and way of life" (ÖVP, 2017, 23).

In 2017, the term then also found its way into the government program of the coalition of the New ÖVP and the FPÖ for the first time, and later again under the New ÖVP and the Greens. In July 2020, the Documentation Center Political Islam was founded, which presented its first report in December of the same year, where it defines it as an "ideology of domination." Lorenzo Vidino (Bridge Initiative Team, 2020a), who wrote several research papers of the Documentation Center Political Islam, repeatedly stated very clearly that "political Islam or legalistic Islamism is more dangerous" (Meier, 2021) than militant jihadism and, conversely, forms the breeding ground for the latter.

This proclaims a state of emergency that must be countered. In this sense, the integration minister of the New Austrian People's Party (ÖVP) also said that political Islam is "poison for our social coexistence and must be fought with all means" (Krone, 2020). Specifically, laws as well as political measures that claimed to legitimize the fight against the so-called political Islam referred to a ban on headscarves in educational institutions, the closure of mosques, and criticism of the political work presented here as racist (Küçükgöl, 2019). At the same time, the representatives of the ÖVP government always pretended to distinguish between Islam as a religion and political Islam. The measures taken by the ÖVP to fight political Islam as described above were judged to be unlawful, be it the headscarf ban that was rescinded by the Constitutional Court (VfGH, 2020), or the ban of the mosques that were rescinded by Administrative Courts (ORF Religion, 2019).

The Documentation Center Political Islam introduced an important catchword to define the notion of "political Islam." According to the institution, it is "an ideology of domination *Herrschaftsideologie* that attempts to influence and shape a state, society, and politics on the basis of values [...] that are against human rights, that is against democracy, and that are also against our constitutional state" (Dokumentationsstelle Politischer Islam, 2022).

In his op-ed in Die Presse, Khorchide, who is head of the Documentation Center's advisory board, stated: "Haven't we already fallen for Political Islam?" He further argued that actors of political Islam would present this term as "a neologism of the Austrian government to attempt to repress the confrontation with Political Islam" and as a strategy aimed to "establishing a victim status for Muslims and second, immunizing themselves from any form of criticism" (Khorchide, 2020). It was this article that was first replicated by me and then sparked a larger debate. In doing so, Arno Tausch replicated my commentary, suggesting that my critique of racism would make an important debate about Islamism and terrorism impossible. It is this important aspect and intervention that I would like to discuss in article.

The self-proclaimed spokesman of neoliberalism, Christian Ortner, who is a regular commentator for Die Presse, then refers to a comparative study on Muslim attitudes. He argues that widespread attitudes among Muslims constitute "a biotope" that "in any case makes it easier for those prepared to use violence to overcome certain civilizational barriers and reach for a weapon or an explosive belt" (Ortner, 2020). He, therefore, calls for the "gray area between fundamentalism, Islamism and terrorism to be illuminated" in order, on the one hand, not to ban opinions—in the spirit of libertarianism—but, on the other hand, still to be able to act against them if they are aiding and abetting the use of violence under criminal law.

In another op-ed, Karl-Peter Schwarz refers to Jan Assmann's concept of "total religion," which would also have applied to Judaism and Christianity for a time. In the sense of catching up with the Enlightenment, Islam as a religion is assumed to have no interpretation, in contrast to the other two monotheistic religions, but to understand the religious sources as literal and to follow them without exception, due to the assumption that they are the word of God. An important argument of Schwarz's is the assertion that in the history of Muslims there has never been "an autonomy of the state vis-à-vis religion." Although "in their history [. . .] Muslims have never fundamentally questioned the totalitarian claim of their religion, they have not always radically brought it to bear either" (Schwarz, 2020). Thus, political Islam is in practice also Islam, and its acceptance is measured by the seriousness with which Muslims devote themselves to their understanding of religion.

3 Counter Position: Political Islam as a Distraction

Rami Ali, Benjamin Opratko, and I each wrote a guest commentary. Ali, a Ph.D.-student of political science, accuses the ÖVP of systematically using the fuzzy term "political Islam" to instrumentalize it "when one has no answers to current challenges and also lacks the necessary will and competence to differentiate between phenomena" (Ali, 2020). He further criticizes the representation that political Islam is more dangerous than jihadism, as claimed by several lead members of the Academic Advisory Board of the Documentation Center Political Islam. At the same time, he also advocates monitoring the Muslim Brotherhood as well as reactionary and anti-democratic movements.

Political scientist Benjamin Opratko, who has published his Ph.D. thesis on anti-Muslim racism in Austria, has primarily countered Khorchide's claim that the terms Islamophobia and anti-Muslim racism have increasingly become combat terms of political Islam. He makes two arguments against Khorchide's definition of political Islam. First, he substantively criticizes that his negative definition, according to which political Islam is opposed to democracy and human rights, does not apply to all Islamic political movements. Second, the institutional-political criticism is aimed at the place of knowledge production. For the Documentation Center Political Islam is "not simply a scientific institute, but advises the state and authorities," (Opratko, 2020) which sometimes entails surveillance and police operations consequently.

My criticism was mainly directed against the use of the term "political Islam" by the Sebastian Kurz-led government, because I saw it as an attack on freedom of religion, freedom of assembly (through the right of association), and freedom of expression. An example of the debate about Islamic kindergartens shows this attitude. At the end of a long politicization of the so-called Islamic kindergartens, by Sebastian Kurz said what he intended: "There should be no Islamic kindergartens" (Bachner, 2017). Summarized in my words at the time, "Muslims should be denied those rights that belong to other religious groups, such as the operation of educational institutions" (Hafez, 2020). And one could add that Muslims should be denied the ability to nurture their children in a religious way like the many Catholic kindergartens in Austria. Overall, then, the criticisms range from using the concept of political Islam as a red herring to steering institutions endowed with power resources, which in turn can be used against critics disagreeable to those in power.

In part, the positions seem to be very far apart, if one considers, for example, the remarks of Khorchide and Schwarz, on the one hand, and the criticism of anti-Muslim racism by state authorities, on the other. At the same time, however, some authors such as Tausch, Ortner, Opratko, and Ali take positions that—at least in theory—do not completely question the concerns of the other side and (keyword freedom of speech) recognize the dangers of government discourse. But where would be the middle ground that, on the one hand, criticizes (also state-institutionalized) racism and, on the other hand, does not deny the phenomenon of political Islam per se, but first takes it and finds a reasonable answer to it? This is the subject of the next subchapter, with a view to Arno Tausch's interjections in his guest commentary.

4 A Problematization of Political Islam with Simultaneous Criticism of Racism

This brings us to an interesting question raised less seriously by Arno Tausch. Can Islamist—not only Islam-marked—actors also be Islamophobic? Scholars like Khorchide and others, whom I have repeatedly criticized in my Islamophobia reports for unreservedly supporting the Islam policy of Sebastian Kurz's New Austrian People's Party, have rejected this criticism as denunciatory with the reference to the fact that they would preach a liberal Islam and criticize political Islam. But in fact, my critique was never about the theological positions of somebody like Khorchide or other of his associates.

This was in fact never the focal point of any of my criticism, and thus this argument seems to be rather an attempt to circumvent my criticism by staging themselves as victims. Yes, this victim discourse is very explicitly pursued and theologically underpinned because my critique was no longer recognized as a scientific-political critique but reinterpreted as a theological critique, presenting it as if Muslims would have a debate amongst others. But that was never the point. Rather, my critique of somebody's support for the government's discriminatory legislation was about the fact that this happened first and foremost in an asymmetrical power relationship. This was the case with Khorchide's unambiguous support of the government's headscarf ban mentioned above as well as the mosque closures. Anyone who has studied the role of the native informant in colonial contexts

(Spivak, 1999), the house and court Jew or Jewish self-hatred in anti-Semitism (Lessing, 2021), or the house n**** in the context of anti-black racism (Malcolm, 1965) outside of Islamophobia research is aware of the possibility of these constellations, where yes indeed, also Muslims can be Islamophobic. Yasemin Shooman uses the notion of "Kronzeuge" (chief witness) in her German texts on the role of Muslims reproducing Islamophobic discourses (Shooman, 2014, 110).

The critique of anti-Muslim racism, however, by far does not only concern those actors who are very consciously supporting the powerful in their discriminatory policies against the rest of the Muslim people. It also affects those Muslims who reproduce Islamophobic thought patterns because of their uncritical attitude and perhaps even out of self-protection. In fact, in my Islamophobia report on Austria, I also criticized those actions on the part of Muslim actors who tend to be on the losing side of Islamophobia. However, this also shows that the research and critical discussion on Islamophobia is not about an imagined "us" versus "them" (the "West" vs. "Muslims") but essentially about a critique of power structures.

Thus, based on my critique of the Documentation Center Political Islam, I critically noted that the Islamic Faith Community in Austria (Islamische Glaubensgemeinschaft in Österreich, IGGÖ) had proposed itself as a cooperation partner of the state authorities (Hafez, 2021, 94). If such cooperation were to happen, it is to be feared that this would not so much lead to more justice being done as that the IGGÖ could exploit its privileged position to strengthen itself as the good Muslim against other Muslim actors on whom it wants to hang the mantle of the bad Muslim. For example, it is worth recalling that the unlawful mosque closures during the ÖVP-FPÖ coalition were supported by the IGGÖ (Hafez, 2019a, 111). I have also shown how thoughtless statements such as those made by the short-lived president of the IGGÖ—who represented the Turkish Ministry of Religious Affairs-related ATIB)—have reproduced Islamophobic discourse positions (Hafez, 2019b, 73).

5 Problematizing Political Islam

Nevertheless, the question remains as to the status of the so-called political Islam. Political Islam is not a phantasm. It is a reality. Apart from the use of the term "political Islam" by Sebastian Kurz's Austrian People's Party, there has been a decades-long debate about this phenomenon on the part of academics as well as political leaders (Hashemi, 2021; Tausch, 2022). Arno Tausch has addressed the phenomenon of political Islam in many of his publications. For example, from the perspective of world systems theory and with quantitative data from World Values Perspectives together with Leonid Grinin and Andrey Korotayev (Leonid et al., 2018) and on the Gulf States (Tausch, 2021b) or related to individual case studies such as on the Muslim Brotherhood (Solomon & Tausch, 2020) or on Islamic Calvinism (Tausch et al., 2010).

In many academic writings, one sees that the concept of political Islam is much broader and serves less to name individual actors. For example, the Oxford Handbook of Islam and Politics (2013), edited by John Esposito and Emad Shahin, includes not only treatises on different parties that refer to Islam and have taken on governmental activities, but also a section on "Jihadi Political Islam," which deals with violent movements such as the Al-Jihad groups in Egypt or Al-Qaeda (2013, 603–642). With this example I want to illustrate that the term political Islam is often used in academic discussions as an umbrella term of a broader phenomenon and not always an analytical category of a very limited phenomenon.

Austria's Muslims, like those outside the country, are just as influenced by various Islamic movements that have originated in the countries of their origin. Yes, organized Muslims in Austria come from milieus where they had joined Sufi brotherhoods and other movements and maintained some sort of relationship with these movements during immigration. Thus, the Turkish Millî Görüş, which is now stronger in Europe than in Turkey itself, is represented in Austria with its offshoots, as are other Islamic political movements. And people who come from the Arab region have not abandoned their connection to Salafiyya, which was actively exported for decades by the Kingdom of Saudi Arabia in the form of Wahhabism, or their socialization in the Muslim Brotherhood once they crossed the border. But what exactly do these references mean in a new environment in which their thinking about politics and religion had to be in a new context? It is precisely at this pivotal point that the question of the state of organized Islam in Europe really needs to be raised.

Let us take one of the most significant movements in Austria with its large Turkish portion of Muslims. Here, the Millî Görüş movement is maybe one of the most typical representations of what is called political Islam in academic literature. Since 2011, the president of the Islamic Religious Community in Austria IGGÖ, which is the legally recognized religious society of Muslims, has been twice from those movements, who have their origin in this tradition. Anthropologists like Werner Schiffauer argue in their research on the post-Islamist turn of the Islamic Community Millî Görüş (IGMG) in Germany drawing on the works of Asef Bayat (2007) that post-Islamism is an intellectual effort to find answers that do not confront the West as the constitutive other as it was the case with the Islamist generation of the founding fathers of Erbakan's Islamism (Schiffauer, 2010, 15). Sociologies Bayat came up with the notion of post-Islamism in his analyses of Islamic political parties in Muslim majority countries, describing those parties that have a religious reference as their source of inspiration, but embrace the idea of a civic state instead of the ideological goal of an Islamic state. In his works he defines post-Islamism as an endeavor to fuse religiosity and rights, faith and freedom, Islam and liberty.

It is an attempt to turn the underlying principles of Islamism on their head by emphasizing rights instead of duties, plurality in place of a singular authoritative voice, historicity rather than fixed scriptures, and the future instead of the past. It wants to marry Islam with individual choice and freedom (albeit at varying degrees), with democracy and modernity, to achieve what some have termed an alter modernity. Post-Islamism is expressed in acknowledging secular exigencies, in freedom from rigidity, in breaking down the monopoly of religious truth. I concluded that whereas Islamism is defined by the fusion of religion and responsibility, post-Islamism emphasizes religiosity and rights. Yet, while it favors a civil and nonreligious state, it accords an active role for religion in the public sphere (Bayat, 2013, 8).

For Bayat, these conservative religious parties denote the end of classical Islamism and represent a new version of a worldview, where its adherents emphasize religiosity, individual choice, and human rights, as well as plurality in place of a singular authoritative Islamist voice (Ibid., 11). Schiffauer is one of the few that has adopted this concept to discuss Muslim organizations in Europe. Schiffauer describes the young leaders of the then-IGMG as post-Islamist organic intellectuals in the rationalization process of Islamism as a revolutionary movement and internal debates of the IGMG (2010, 30–33).

Like Bayat, for him, "post-Islamism is an attempt to relate Islam to the individual rights, majority and multivocality, openness rather than closure, and future instead of past" (2010, 17). Post-Islamism is a form of "alternative modernity." For Schiffauer, the post-Islamist generation represented a generation that embraced the German language as well as the German polity (2010, 371), but also legally contested the work of the intelligence agency, the Federal Office for the Protection of the Constitution (Bundesamt für Verfassungsschutz). The latter argued that the IGMG was a manifestation of legalistic Islamism, thus its post-Islamist turn was not believed to be authentic. There is no similar literature like Schiffauer's on the situation in Austria.

Similarly, there is a debate on the most prominent two cases of the phenomenon of post-Islamism in the wake of the Arab Spring, one being Egypt and the other being Tunisia. While in Tunisia, the Ennahda party seems to perfectly represent the idea of a successful political party that has embraced democracy and while not neglecting its ideological roots that are traced back to the Muslim Brotherhood, they clearly embrace free elections, political pluralism, and peaceful alteration of power and programmatically leave "political Islam" to embrace "Muslim democracy" (Ghannouchi, 2016), while the Egyptian Freedom and Justice Party as the political arm of the Muslim Brotherhood was "driven by a hunger for power fueled by the immense vacuum left after the fall of Mubarak" (Al-Anani, 2015, 530). As Ben Lazreg sums up in his comparative analysis of both cases, "Unlike the Ennahda party in Tunisia, which endorsed a spirit of reconciliation, consensus, compromises, and coalition in order to safely manage the transitional process, the MB was rigid, exclusive, and unable to adapt and handle the fast-changing transitional period" (Ben Lazreg, 2021, 11).

Following a report on the Millî Görüş published by the Documentation Center Political Islam (Dokumentationsstelle Politischer Islam, 2021), the Austrian Islamic Federations, which are part of the IGMG with its headquarters in Cologne, Germany, published a response (IFÖ, 2021). The Islamic Federations are a good example of the role of political Islam in the sense of Islamic movements participating in democratic orders, because they themselves see their historical roots in the father of political Islam in Turkey. The Islamic Federations' self-portrayal states:

> Due to their basic religious orientation, the founding generation was close to Necmettin Erbakan, the founder of the Millî-Görüş movement and thus also the chairman of the Millî-Görüş parties in Turkey. Over the years, this closeness developed into an exchange with mutual support (IFÖ, 2021, 8–9).

At the same time, the Islamic Federations criticize, for example, that the basic report of the Documentation Center does not focus on the concrete organization on the ground in Austria as the object of research but would draw conclusions about the organization in Austria with the help of references to the Millî Görüş movement in Turkey (IFÖ, 2021, 6–7). For example, they point out that the equation "Necmettin Erbakan = Millî Görüş = IGMG = Islamic Federations" (IFÖ, 2021, 11) is inadmissible. Thus, claims by the Documentation Center that the Islamic Federation are representatives of the Millî Görüş movement are countered with an interview excerpt from the chairman of the IGMG, who to declare that his IGMG, based in Cologne, would be "structurally, formally and financially independent" (Ibid.) for the avoidance of doubt. Moreover, it is stated "that the paths of the IGMG, and with it those of the Islamic Federations, parted ways with the Millî-Görüş ideology decades ago" (Ibid.). Taking the example of anti-Semitism, the founder of the Millî Görüş movement and the leader of the Islamic parties in Turkey, Necmettin Erbakan, has been reproached (Bali, 1999), although in the Turkish context not Erbakan alone (Bali, 2012). Thus, the Islamic Federations criticize:

> Anyone looking in the 'Report' (of the Documentation Center, F.H.) for the substantive differences of the Islamic Federations with Erbakan's politics or the Millî-Görüş movement, which have already been described in this statement, will not find anything, especially with regard to religious pluralism, homophobia or anti-Semitism (IFÖ, 2021, 19).

And later, the Islamic Federations clearly declare their opposition to anti-Semitism, claiming that they "regard every form of racism as hostility against humans (Menschenfeindlichkeit) and fight excesses of every kind, including explicitly anti-Semitism, with all determination" (IFÖ, 2021, 20). They go on to state, even more importantly for the case of Austria, that anti-Semitism led to the Holocaust. And further: "Learning from this history and keeping the culture of remembrance alive is a common responsibility and at the same time an important reminder" (IFÖ, 2021, 20).

Three important remarks are to be made at this point: First, the Islamic Federations clearly distance themselves from the political leadership of classical political Islam in Turkey in the twentieth century and explicitly note that, in their view, differences exist in connection with pluralism, homophobia, and anti-Semitism. Second, they clearly oppose anti-Semitism. And third, they also link their fight against anti-Semitism to a culture of remembrance, which is significant in a country like Austria or Germany as historical perpetrator nations in Nazism.

Regardless of the evaluation of the Documentation Center's criticism of the Islamic Federations, this threefold commitment of the Islamic Federations means, on the one hand, that they are documenting a change to which they also acknowledge in their self-representation when they say that the paths of the Millî Görüş and of the

IGMG and the Islamic Federations diverged decades ago, when the followers of the Islamic Federations had made Austria a center of their lives.

As an essential part of the Islamic Faith Community IGGÖ, the Islamic Federations have also been part of various processes such as the European Imams Conferences, which took place in 2003, 2007, and 2010 and took very clear positions regarding social issues on the role of Muslims in Europe (Hafez, 2013). For example, the Graz Declaration of 2003 states:

> The medieval division into a world of opposites of Dar as Islam = House of Islam and Dar al harb = House of War is to be rejected. It has no basis in the Qur'an, nor in the Sunnah, and as a historical, long-outdated phenomenon, is of no relevance today (EIC, 2003).

Moreover, there is an explicit affirmation of human rights as a "central component of Islam" (Ibid.). While the 2003 conference still speaks of men and women being "equal partners to each other" (Ibid.), the 2010 final declaration additionally refers to the "Quranic principles of equality and partnership of men and women" (EIC, 2010), which is to be "implemented in associations and mosque communities" (Ibid.). The statement argues that "society must not deprive itself of half of its potential [. . .] by cutting women off from opportunities for participation" (Ibid.). Building on this, it calls for a "renewed scrutinizing looks at many fiqh (legal) regulations [. . .] concerning women and their role and position" (Ibid.).

This demand is found among several other demands in areas relating to bodies of organizations, education, dialogue, legal protection, and more. It would be untrustworthy if imams of the Islamic Federations in Austria were to endorse these declarations without them being in any way relevant to their own organization. At the same time, these proclamations are a yardstick to evaluate the reality of these organizations. When it comes to the role of women for instance, their participation within the overall leadership positions is still irrelevant, as it was shown already in one of my earliest studies on the IGGÖ in 2006 (Hafez, 2006).

Even if the final declarations of the Imam Conferences were supported by the Islamic Federations, they are not the same as the positions of the latter. Compared with other successful post-Islamist parties, such as Ennahda in Tunisia, one thing can be said here: In Austria, Muslim associations such as the Islamic Federations lack programmatic as well as theoretical confrontations with their own roots, history and present. On the one hand, this is understandable against the background that these associations were largely built up by the guest worker generation. In the meantime, however, the leadership cadre has been rejuvenated, and the rank-and-file members of these associations also include academics, who are increasingly expected to engage in a self-critical debate.

6 Conclusion

In this respect, two opposing positions can be observed today: On the one hand, critics assume that associations such as the Islamic Federations in Austria are nothing more than offshoots of the Millî Görüş movement in Austria. A continuity of people and ideology is assumed, which leads to the positioning of associations such as the Islamic Federations here being regarded as not credible once they distance themselves from parts of their spiritual founder's legacy such as Necmettin Erbakan in this case.

The other position is to take the Islamic Federations at their word and assume that associations such as the Islamic Federations, which have their ideological roots in the Millî Görüş movement, have evolved, much in the same way that social democratic parties today do not represent an orthodox representation of Karl Marx and Friedrich Engels or have renounced violence as an instrument of political power as has the ÖVP, which distances itself from the Austro-Fascist regime that ruled Austria between 1934–1938, while still having troubles with coming to terms with its own historical remembrance of the Austro-Fascist leader Engelbert Dollfuß (Hagen, 2021).

In other words, the post-Islamist turn is not recognized. In this respect, there is no need for a recurring distancing from the historical roots from which these associations have already distanced themselves. As the Islamic Federations also show in their reply to the report of the Documentation Center, this report of the Documentation Center also did not examine any contemporary publications, press releases, and the like by the Islamic Federations that would contradict the image of classical political Islam of the mid-twentieth century. At the same time, however, Muslim associations such as the Islamic Federations lag behind an ideological rationalization of their day-to-day political positions and demand even more the production of their own writings, which provide information about the reception of their own history as well as their views of contemporary issues.

As this article has also shown, representatives of Muslim institutions are as much involved in the reproduction of Islamophobic structures when it comes to defending and stabilizing their privileges and positions of power. The Islamic Religious Community in Austria (IGGÖ) that initiated the closure of several mosques, which the Austrian government later presented as a fight against political Islam, is a great example for this. And this shows once more, as Samir Amin has once said with a critical tone toward Islam that "Islam, like Christianity, can adapt itself to everything" (Amin, 1977, 95).

Hence, analyzing and criticizing institutions of power ultimately goes hand in hand with a critique of racism, as the latter is ultimately a critique of power structures. Samir Amin had once criticized political Islam for being at the "service of imperialism" (Amin, 2007, 3). While this argument has prompted critical response (Amin-Khan, 2009), there is still some truth in that throughout history, most Muslim clerics have tended to ally themselves with those in power (Hafez, 2014) and inquiring relationships of power between Muslim religion and politics seems to be an important aspect of this discussion that is too often neglected.

References

Al-Anani, K. (2015). Upended path: The rise and fall of Egypt's Muslim Brotherhood. *Middle East Journal, 69*(2), 527–543.

Ali, R. (2020). *Terrorismus in Wien: Fatale Schuldabwehr*. Die Presse. 13 Nov 2020. Retrieved August 4, 2022, from https://www.diepresse.com/5896904/terrorismus-in-wien-fatale-schuldabwehr?from=rss

Amin, S. (1977). Imperialism and unequal development. *Monthly Review Press, 19*(6), 2–17.

Amin, S. (2007). Political Islam in the service of imperialism. *Monthly Review, 59*(7), 1–19.

Amin-Khan, T. (2009). Analysing Political Islam: A critique of traditional historical materialist analytic. *Monthly Review Press, 61*(4), 7–10.

Bachner, M. (2017). *Kurz im KURIER-Gespräch: "Islamische Kindergärten abschaffen."* Kurier. 21 Jun 2017. Retrieved August 4, 2022, from https://kurier.at/politik/inland/sebastian-kurz-im-kurier-gespraech-islamische-kindergaerten-abschaffen/271.008.503

Bali, R. N. (1999). The image of the Jew in the rhetoric of political Islam in Turkey. *Cahiers d'études sur la Méditerranée orientale et le monde turco-iranien, 28*(1), 1–11.

Bali, R.N. (2012). Model citizens of the state: The Jews of Turkey during the multi-party period. .

Bayat, A. (2007). *Making Islam democratic: Social movements and the post-Islamist turn*. Stanford University Press.

Bayat, A. (2013). *Post-Islamism: The changing faces of political Islam*. Oxford University Press.

Ben Lazreg, H. (2021). Post-Islamism in Tunisia and Egypt: Contradictory trajectories. *Religions, 12*(4), 1–22.

BKA. (2020). *Bundeskanzler Kurz: Terrorismus und politischen Islam mit allen Mitteln bekämpfen.* 11 Nov 2020. Retrieved August 4, 2022, from https://www.bundeskanzleramt.gv.at/bundeskanzleramt/nachrichten-der-bundesregierung/2020/bundeskanzler-kurz-terrorismus-und-politischen-islam-mit-allen-mitteln-bekaempfen.html

Bridge Initiative Team. (2020a). *Factsheet: Lorenzo Vidino*. 22 Apr 2020. Retrieved August 4, 2022, from https://bridge.georgetown.edu/research/factsheet-lorenzo-vidino/

Bridge Initiative Team. (2020b). *Factsheet: Sebastian Kurz*. 17 Apr 2020. Retrieved August 4, 2022, from https://bridge.georgetown.edu/research/factsheet-sebastian-kurz/

Bridge Initiative Team. (2021). *Factsheet: Documentation Center Political Islam.* 12 Dec 2021. Retrieved August 4, 2022, from https://bridge.georgetown.edu/research/factsheet-documentation-center-political-islam-dokumentationsstelle-politischer-islam/

Der Standard. (2022). *'Blattlinie' Der Standard*. Retrieved August 4, 2022, from https://about.derstandard.at/redaktion/blattlinie/

Die Presse. (2010). Die Presse-Blattlinie. *Die Presse*. 25 Nov. 2010. Retrieved August 4, 2022, from https://www.diepresse.com/613276/die-presse-blattlinie

Die Presse. (2020). *Die heftige Debatte um den "Politischen Islam"* 25 Nov. 2020. Retrieved August 4, 2022, from https://www.diepresse.com/613276/die-presse-blattlinie

Dokumentationsstelle Politischer Islam. (2021). *Die Millî Görüş Grundlagenbericht der Dokumentationsstelle Politischer Islam.* Grundlagenpapier. Ausgabe Nr. 01/2021 Retrieved August 4, 2022, from https://www.dokumentationsstelle.at/uber-uns/

Dokumentationsstelle Politischer Islam. (2022). *Definition Politischer Islam*. Retrieved August 4, 2022, from https://www.dokumentationsstelle.at/uber-uns/

EIC. (2003). *Grazer Erklärung der Konferenz "Leiter islamischer Zentren und Imame in Europa" Graz am 15. Juni 2003*. European Imam Conference. Retrieved August 4, 2022, from www.al-sakina.de/inhalt/bibliothek/grazer_Erklaerung/grazer_erklaerung.html

EIC. (2010). Schlusserklärung der 3. *Konferenz europäischer Imame und SeelsorgerInnen vom 14. bis 16. Mai in Wien. European Imam Conference*. Retrieved August 4, 2022, from https://web.archive.org/web/20150801073314/http://www.europa-imamekonferenz.com/

Esposito, J. L., & Shahin, E. (Eds.). (2013). *The Oxford handbook of Islam and politics*. Oxford University Press.

Ghannouchi, R. (2016). From political Islam to Muslim democracy: The Ennahda party and the future of Tunisia. *Foreign Affairs, 95*(7), 58–75.

Grinin, L., Korotayev, A., & Tausch, A. (2018). *Islamism, Arab Spring, and the future of democracy: World system and world values perspectives. Series: Perspectives on Development in the Middle East and North Africa (MENA) Region*. Springer.

Hafez, F. (2006). *Die Islamische Glaubensgemeinschaft in Österreich: Eine Analyse der Organisationsstruktur unter besonderer Berücksichtigung der Rolle muslimischer Spitzenverbände*. Master Thesis, Univeristy of Vienna.

Hafez, F. (2013). Muslimische Aggiornamenti und Denominationalismus in Österreich. Eine Analyse des Fallbeispiels Islamische Glaubensgemeinschaft in Österreich. In J. Nautz, K. Stöckl, & R. Siebenrock (Eds.), *Öffentliche Religionen in Österreich. Politikverständnis und zivilgesellschaftliches Engagement* (pp. 211–224). Innsbruck University Press.

Hafez, F. (2014). *Islamisch-politische Denker: eine Einführung in die islamisch-politische Ideengeschichte*. Peter Lang Publishing.

Hafez, F. (2019a). Islamophobia in Austria: National Report 2018. In E. Bayraklı & F. Hafez (Eds.), *European Islamophobia Report 2018* (pp. 87–125). SETA.

Hafez, F. (2019b). *Österreichischer Islamophobieberich*. SETA.

Hafez, F. (2020). Die Verteidigung der Grundrechte ist oberstes Gebot der Stunde. *Die Presse*. 19 November 2020. Retrieved August 4, 2022, from https://www.diepresse.com/5899979/die-verteidigung-der-grundrechte-ist-oberstes-gebot-der-stunde?from=rss

Hafez, F. (2021). Islamophobia in Austria: National Report 2020. In E. Bayraklı & F. Hafez (Eds.), *European Islamophobia Report 2020* (pp. 81–121). Leopold Weiss Institute.

Hafez, F. (2022). *Das Dispositiv ,Politischer Islam' in der österreichischen Bundespolitik*. Frankfurter Zeitschrift für Islamisch-Theologische Studien.

Hagen, L. (2021). 'Der designierte Innenminister und das Dollfuß-Museum in seiner Heimat.' *Der Standard*. 5 Dec 2021. Retrieved August 4, 2022, from https://www.derstandard.at/story/2000131678077/der-designierte-innenminister-und-das-dollfuss-museum-in-seiner-heimat?ref=article

Hashemi, N. (2021). Political Islam: A 40 year retrospective. *Religions, 12*(2), 1–17.

Islamische Föderation in Österreich. (2021). *Grundlagenbericht der "Dokumentationsstelle Politischer Islam"über, Die Millî Görüş*. Islamische Föderation in Österreich.

Khorchide, M. (2020). Sind wir nicht schon längst auf den Politischen Islam hereingefallen? *Die Presse*. 17 November 2020. Retrieved August 4, 2022, from https://www.diepresse.com/5898835/sind-wir-nicht-schon-laengst-auf-den-politischen-islam-hereingefallen?from=rss

Krone. (2020). Dokumentationsstelle nimmt Extremismus ins Visier'. *Kronen Zeitung*. 15 July 2020. Retrieved August 4, 2022, from https://www.krone.at/2192166

Küçükgöl, D. (2019). Spricht sie noch oder schweigt sie schon? Silencing-Strategien gegen muslimische Frauen in der feministischen Praxis. In A. M. Heinemann & N. Khakpour (Eds.), *Pädagogik sprechen* (pp. 83–98). JB Metzler.

Lessing, T. (2021). *Jewish self-hate*. Berghahn Books.

Malcolm, X. (1965). *The House Negro and the Field Negro*. Speech. Retrieved August 4, 2022, from https://www.youtube.com/watch?v=7kf7fujM4ag

Meier, C. (2021). Projekt gesellschaftliche Umgestaltung. *FAZ*. 11 February 2021. Retrieved August 4, 2022, from https://www.faz.net/aktuell/feuilleton/debatten/lorenzo-vidino-ueber-den-politischen-islam-in-deutschland-17183411.html

Opratko, B. (2020). Islamophobie und politischer Islam: Worum es im Konflikt geht. *Die Presse*. 24 November 2020. Retrieved August 4, 2022, from https://www.diepresse.com/5902147/islamophobie-und-politischer-islam-worum-es-im-konflikt-geht?from=rss

ORF Religion. (2019). *Moscheenschließungen rechtswidrig – Kultusamt beruft*. 14 February 2019. Retrieved August 4, 2022, from https://religion.orf.at/v3/stories/2964551/

Ortner, C. (2020). Der islamische Fundamentalismus als Helfer des Terrors. *Die Presse*. 19 November 2020. Retrieved August 4, 2022, from https://www.diepresse.com/5900010/der-islamische-fundamentalismus-als-helfer-des-terrors?from=rss

ÖVP. (2017). *Der neue Weg. Ordnung & Sicherheit 3/3.* Das Programm der Liste Sebastian Kurz – die neue Volkspartei zur Nationalratswahl. OVP.

Schiffauer, W. (2010). *Nach dem Islamismus. Eine Ethnographie der Islamischen Glaubensgemeinschaft Milli Görüş*. Suhrkamp.

Schwarz, K. P. (2020). Wer den, totalen Islam will, hat in Österreich nichts verloren. *Die Presse*. 24 November 2020. Retrieved August 4, 2022, from https://www.diepresse.com/5902284/wer-den-totalen-islam-will-hat-in-oesterreich-nichts-verloren?from=rss

Seeh, M. (2022). Beschwerdeflut nach Anti-Terror-Razzia. *Die Presse*. 2 February 2022. Retrieved August 4, 2022, from https://www.diepresse.com/5931839/beschwerdeflut-nach-anti-terror-razzia

Shooman, Y. (2014). *Weil ihre Kultur so ist «: Narrative des antimuslimischen Rassismus. Transcript.* Bielefeld Verlag.

Solomon, H., & Tausch, A. (2020). The Demise of the Muslim Brotherhood in the Arab World? *Jewish Political Studies Review, 31*(1/2), 171–209.

Spivak, G. C. (1999). *A critique of postcolonial reason: Toward a history of the vanishing present.* Harvard University Press.

Styria Media Group. (n.d.). *'Die Presse'*. Retrieved August 4, 2022, from https://www.styria.com/en/brands/die-presse

Tausch, A. (2021a). From the periphery to the center of global knowledge production? A bibliometric analysis of the evolution of a social science community from a small country: Austria. *Journal of Globalization Studies., 12*(2), 69–102.

Tausch, A. (2021b). *The future of the Gulf region: Value change and global cycles*. Springer.

Tausch, A. (2021c). Social attitudes fueling Islamist terrorism. In H. Solomon (Ed.), *Directions in international terrorism* (pp. 187–218). Springer.

Tausch, A. (2021d). Islamist terrorism, political Islam and migration in Western Europe. In H. Solomon (Ed.), *Directions in international terrorism* (pp. 289–334). Springer.

Tausch, A. (2022). Political Islam and religiously motivated political extremism. An international comparison. .

Tausch, A., Bischof, C., & Mueller, K. (2010). *Muslim Calvinism. Internal security and the Lisbon process in Europe. European diversity series*. Rozenberg Publishers.

VfGH. (2020). *Verhüllungsverbot an Volksschulen ist verfassungswidrig*. 11 December 2020. Retrieved August 4, 2022, from https://www.vfgh.gv.at/medien/Verhuellungsverbot_an_Volksschulen_ist_verfassungswid.de.php, (Access date 15 Jul 2021).

Part IV
Security

States of Emergency: In Whose Interest Are They Invoked?

José Alemán, Dong Wook Lee, and Dwayne Woods

Abstract Arno Tausch's illustrious career spanned five very prolific decades. The topics he explored include political stability, militarization, terrorism, and political violence, to name a few. In this essay, we honor his legacy with research that investigates why governments declare a state of emergency. Scholars see in these events instances of political instability. We argue that emergencies can create potential opportunities for executives to reshape the balance of power in their favor. However, the opportunity dynamics they open, either through extra-constitutional means (e.g., a coup or auto-coup) or autocratization, vary by regime type and are contingent on the personalization of executive power within a regime. In tending to this heterogeneity in regimes' invocations of emergencies, we follow Arno Tausch's exemplary use of quantitative analysis to answer complicated questions. Arno Tausch relied on both deductive and inductive methodologies. We first use an inductive approach to uncover meaningful patterns. We then conduct a nonlinear analysis of state-of-emergency (SOE) episodes from 1950 to 2010. Our findings indicate that multiparty dictatorships produce significantly more SOEs compared to democracies. However, they witness fewer SOEs in the presence of more powerful executives. The latter are also more likely to trigger an autocratization episode.

J. Alemán (✉)
Department of Political Science, Fordham University, The Bronx, NY, USA
e-mail: aleman@fordham.edu

D. W. Lee
Department of Political Science, Adelphi University, New York, USA
e-mail: dwlee@adelphi.edu

D. Woods
Department of Political Science, Purdue University, West Lafayette, IN, USA
e-mail: dwoods2@purdue.edu

G. Segell (ed.), *Development, Globalization, Global Values, and Security*,
https://doi.org/10.1007/978-3-031-24513-8_13

1 Introduction

On October 10th, 1972, South Korea's president, former army general Park Chung-hee, declared martial law, suspending the constitution and dissolving the National Assembly. Park proclaimed a new constitution, Yushin (or revitalization), giving him sweeping executive and legislative powers, the right to appoint one-third of all seats in the country's unicameral legislature and all Supreme Court justices (Chang, 2015, 42), and the ability to face an indirect election by an electoral college with no limits on reelection.

Once ratified in a referendum, the Yushin constitution effectively turned Park's self-coup into a legal dictatorship. Seven years later, on October 26th, 1979, President Park died at the hands of his own CIA (KCIA) chief. Although Prime Minister Choi Kyu-hah became acting president pursuant to the Yushin Constitution, Major General Chun Doo-hwan declared martial law on May 17th, 1980, installing a military junta and inaugurating a new dictatorship.

The Korean case illustrates how emergency declarations can be used to introduce new autocratic regimes. In Turkey, however, authoritarian power grabs—increased control over the media and the judiciary and intensified repression—were occurring even before emergency rule was declared. The 2016 state of emergency (hereafter called SOE) thus functioned as a mechanism to "consolidate further authoritarian rule" (Lührmann & Rooney, 2020: 1). Solomon and Tausch (2019, 258) thus refer to Erdogan's presidency as "imperial."

In Taiwan, we observe yet another pattern related to the longevity of emergency rule use when it becomes an institutional feature of autocracy. Essentially, the country was under martial law from 1949 to 1987, the "longest unbroken stretch of martial law" in the world (Greitens, 2016: 75) and a period coinciding with the single-party rule of the Nationalist (or Kuomintang) party.

These vignettes prompt a question. What type of opportunity structures do declarations of emergency (i.e., martial law, the state of siege, and the state of emergency) open for regimes? In a Rikerian frame, a pertinent observation is whether states of exception amount to "politics by other means" that fundamentally change the balance of power and institutional rules within the polity. Riker conceived of institutions "as the equilibria that emerge from the strategic behavior of individuals" (Pepinsky, 2014, 263–264). Riker's key insight is that "institutional rules are subject to strategic manipulation" (Pepinsky, 2014, 624). It is not easy to observe such manipulation. However, SOEs afford us the possibility to exploit manifestations of unobservable strategic behavior since these states are often evoked in response to either unforeseen or unplanned shocks.

We theorize that SOEs are endogenous shocks that create political opportunities for some executives to refashion the balance of power in their favor. The theory postulates that the opportunity to change the status quo engendered by these events varies by regime type and how efforts to personalize power interact with these institutional factors. As such, there should be two types of power shifts occurring under emergencies that manifest themselves in the data. The first would be a move toward more personalization of power within regimes, most notably in multiparty regimes where the ruling party enjoys more bargaining power vis-à-vis the executive (Magaloni, 2008, 717). This is a possibility alluded to (Bjørnskov & Voigt, 2018a, 122) but not yet investigated. Secondly, SOEs should create opportunities for autocratization through extra-constitutional means like a coup or an auto-coup.

At different historical junctures, militaries have used crises to seize power. As Tausch warned, "violence and coups d'état are related to the militarization of development and to the strains that this military burden causes for the ability of the government to control the savings rate." (Tausch & Prager, 1993, 89). Viewed through this theoretical lens, we approach heterogeneity in regimes' invocations of emergencies as information signals about the status quo and ways in which SOEs can be used to change the balance of power and in whose favor.

We thus adopt a nonlinear model specification that allows us to estimate the effect of regime type on the power dynamics revealed with the invocation of emergency provisions while controlling for legislative and judicial constraints on the executive. The only other systematic analyses of the determinants of emergency rule, the studies by Bjørnskov and Voigt, do not employ complete data on emergencies. Instead, they use measures of political unrest to proxy for them in years in which their dataset registers no emergency declarations (Bjørnskov & Voigt, 2018a, 115, 2020, 586).

Using panel data on SOE episodes from 1950 to 2010, we find that multiparty dictatorships experience significantly more SOEs compared to other regime types. On the other hand, personalized multiparty regimes (or multiparty autocracies with more powerful executives) witness fewer SOEs and a longer spell in between them. At the same time, their emergencies are more likely to trigger an autocratization episode.

This article is organized into five sections. We begin by placing SOEs in the context of the research program Tausch developed around questions of political stability and violence. We then provide a thorough descriptive examination of the data on these events. Our examination is exploratory since we look to see what type of clusters emerge from the data temporally and spatially. Also, we want to bring into dialogue the evidence we present with Tausch's insights regarding the causes and consequences of militarization, executive turnovers, and political repression. Following this descriptive analysis, we model SOEs by distinguishing between the onset of a state of exception and its continuation since this temporal dimension in the data conveys important information regarding the degree of personalism before and after an SOE is invoked. Next, the findings from our empirical analysis are presented and discussed. We conclude by situating our analysis within a broader literature on state failure, human rights violations, and democratization.

2 What Is an Emergency Declaration?

Emergency measures are usually defined as legal measures that invoke derogations from normal constitutional order (Keith & Poe, 2004, 1075–1076). In principle, they are summoned to facilitate an effective governmental response to a crisis (Richards & Clay, 2012). However, such derogations can entail suspensions or restrictions of fundamental rights (Richards & Clay, 2012), expansion of executive authority vis-à-vis other branches (Bjørnskov & Voigt, 2018a, 2018b), and creation of special administrative or judicial mechanisms. As Honig (2009, 67) points out, emergency declarations entail the "redistribution of the existing powers of governance from procedural processes to discretionary decision."

The roots of emergency declarations can be traced back to the Roman Republic and its use by Roman consuls. The consuls accorded a dictator nearly absolute power to act independently of typical republican principles. However, its contemporary understanding and link to regime and constitutional legitimacy emerged during the French Revolution. It was with the decree of an SOE by the French Constituent Assembly that the idea of exceptional circumstances granting a regime the legitimacy to suspend normal constitutional guarantees arose.

Basically, the dichotomy arose in which "a state of peace" differed from a "state of siege," "where under the latter, all the functions entrusted to civilian authority for maintaining order and internal policing pass to the military commander, who exercises them under his exclusive responsibility" (Sheeran, 2013, 496). In the nineteenth century, the idea of exceptional powers gained a constitutional footing with the adoption of these powers by the executive during times of "state of siege" in the 1848 constitution of the Second French Republic and the suspension of habeas corpus at the onset of civil war in the USA.

Militarization became a common phenomenon in the twentieth century as countries in the periphery sought to overcome their dependent status through insertion into the world economy (Tausch & Heshmati, 2012, 41). As Tausch found more than three decades ago, "[i]n the semi-integrated countries of the capitalist periphery and semi-periphery, there is an alarming positive effect of militarism and public spending on growth…The stage was therefore set for a massive militarization of the periphery." (Tausch, 1989a, 648). The first use of a state of exception to ensconce authoritarian rule in a constitutional framework did not occur in a military regime, however. It happened in Nazi Germany, which never abrogated the Weimar Constitution; "instead, Hitler exercised emergency powers pursuant to the Enabling Act [of 1933] through to the end of the war." (Ginsburg & Huq, 2018, 40).

In the post-World War II era, SOEs have been used extensively by actors in different governments. For example, President Truman declared an SOE in response to the conflict in Korea in 1950 (Sheeran, 2013, 298). Charles De Gaulle was granted exceptional powers under an SOE in 1958 to counter a military putsch in Algeria. Other instances followed. "Latin America…provides perhaps the densest [sic] thicket of historical examples [during the autocratic wave of the 1960s and 1970s] of both coups and misused emergency powers" (Ginsburg & Huq, 2018: 52).

A good example is Chile. Upon taking power, the Chilean armed forces declared a state of siege that was "understood as 'a state or time of war.' The state of siege was renewed every six months until March 1978, when it was replaced by a 'state of emergency'" (Policzer, 2009, 187). "From March 11, 1981 . . . [until the transition to democracy in 1990] Pinochet kept the country perpetually under both a state of emergency. . . and a state of danger . . . He also declared a state of siege twice." (Hilbink, 2007, 136).

Some autocratic regimes in the Middle East and North Africa also "maintained state of emergency laws for decades" (Barany, 2016, 134). Take Egypt for example, where an SOE lasted during the entirety of Hosni Mubarak's tenure as president (1981–2011), ending only in May 2012 (Richards & Clay, 2012, 450). South Korea, on the other hand, witnessed twelve garrison and emergency decrees in the 1970s, all under President Park Chung-hee (Greitens, 2016, 161) as examined by Domínguez (2011, 593).

Between 1985 and 2014, "some countries invoked a state of emergency at least once" (Ginsburg & Huq, 2018: 58). Today, "some 90% of constitutions in force" "have some provisions on emergency powers" (Bjørnskov & Voigt, 2018b; Ginsburg & Huq, 2018, 58). However, "emergency provisions often provide little specificity of which actions short of open warfare warrant the declaration of a state of emergency" (Lührmann & Rooney, 2020, 5). The result is that their causes and duration vary.

3 Measuring States of Emergency: Descriptive Analysis

This section explores patterns in the invocation of SOEs and their links, if any, to regimes. For data on emergency declarations, we rely on version 10 of the Varieties of Democracy (V-Dem) project, which includes information on whether an emergency was declared in a specific country-year to cope with various types of events including interstate wars, domestic conflicts, and natural disasters (Coppedge et al., 2020, 213). We only consider emergencies invoked in response to overt domestic conflict in the form of a terrorist attack (Bjørnskov & Voigt, 2020), civil war, or mass protest.

Tausch published "numerous works examining terrorism in its myriad manifestations" (Solomon forthcoming; Tausch forthcoming). We echo his concern because actions like these are often a sign that institutions are not working well or that politics may not be fully institutionalized (Gerring et al., 2021, 610–11). These events, although emanating from below, sometimes reflect or arise in response to elite conflicts and divisions. We thus follow Tausch and Heshmati's (Tausch & Heshmati, 2012, 50) recent call for political scientists to work "with a wider concept of 'social cohesion', not just analyzing income inequality, but also . . . human rights," political exclusion and political repression.

To generate this SOE declaration data, country experts were asked, for each situation, whether an emergency was declared/existed each year or not, resulting in a series of dichotomous scales. These expert coding were then used to generate the "average" country-year values for each emergency type. Because the resulting indicators range from 0 to 1 and various transformations applied to this left-censored and skewed data failed to make it normal, we opted to create a dichotomous variable taking a value of 1 if the underlying indicator's value is greater than 0 or 0 otherwise (Hafner-Burton et al., 2011; Lührmann & Rooney, 2020).

Figure 1 provides a sense of the geographic distribution and intensity of emergency rule by regime type from 1950 to 2010. The lattice graph confirms the ubiquity of states of exception across political regimes.

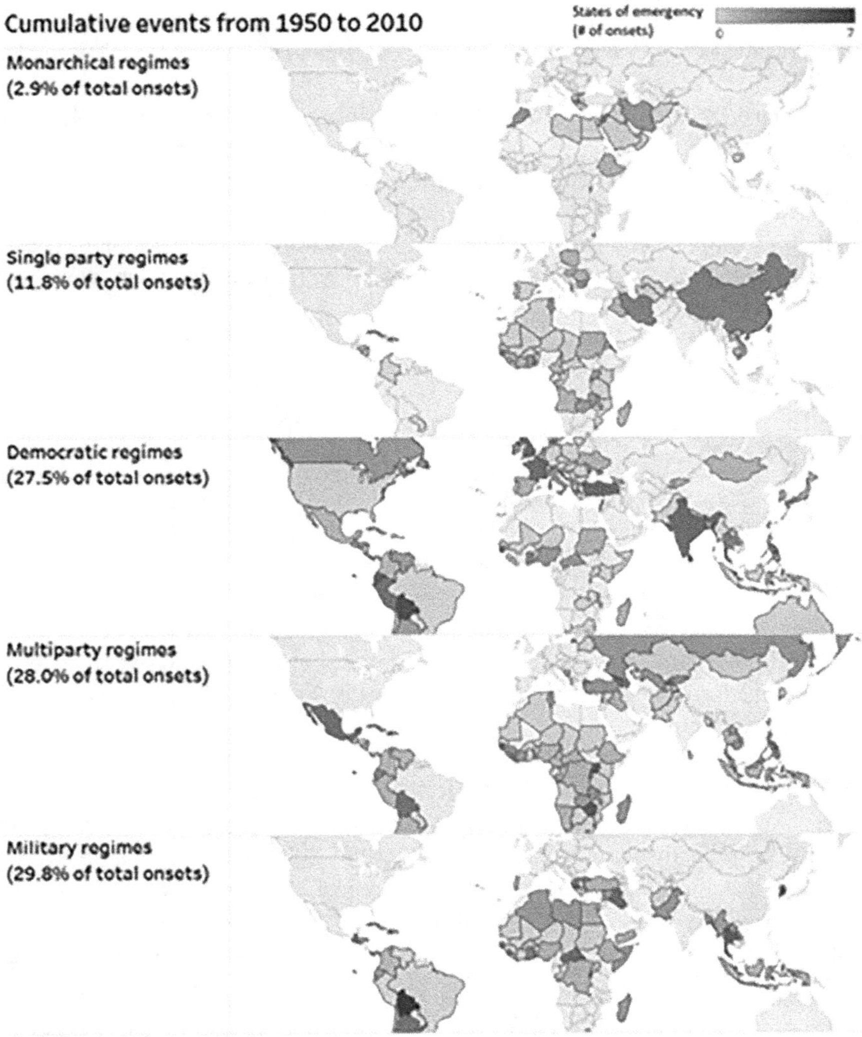

Fig. 1 Spatial clusters of emergencies by regime type

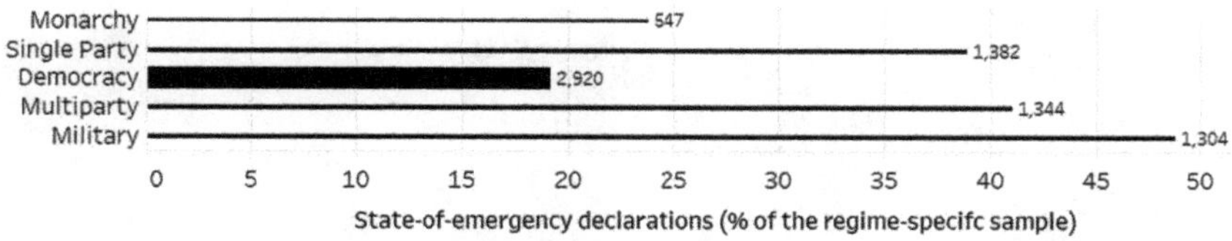

Fig. 2 Weighted proportions of state-of-emergency declarations by regime type (1950–2010)

From the data displayed in Fig. 1, a few stylized facts emerge. First, the average within-country non-emergency spell, or the time between two emergencies, is 9 years. Second, when the frequency by regime type of states of exception (counted as onset) is taken into consideration, monarchies had the lowest incidence as a share of emergencies (2.9% of 346 total episodes), with single-party regimes following behind (11.8%). Military regimes had the highest frequency (29.8%), followed by multiparty autocracies (28.0%) and democracies (27.5%). Except for monarchies then, there is not a major difference in SOE frequency between democracies and nondemocracies.

Similarly, Fig. 2 provides "normalized" frequency data on emergency incidents (i.e., as a proportion of the regime-specific sample because emergencies are more frequently observed in some regimes than others). The figure reveals regime-related heterogeneity in the use of emergencies. First, SOEs are observed more in military, multiparty, and single-party autocracies relative to those in absolute monarchies and democracies. Second, democracies exhibit the lowest probability of experiencing an SOE.

Although democracies more frequently respond to political unrest using emergency powers (Bjørnskov & Voigt, 2018a: 111), the danger for them lies in emergency clauses moving some toward autocracy or hybrid regimes (Lührmann & Rooney, 2020). This democratic backsliding can occur through a self-coup (Lührmann & Lindberg, 2019), an alliance between the executive and the military to dissolve parliament (Pérez-Liñán, 2007, 50).

Our preliminary look at the data reveals that as early as the 1980s, Tausch foreshadowed some of the patterns we note. Writing about arms exports from the Soviet Union and its satellites to the Third World, he observed that "a considerable degree of political instability, as measured by the frequency of illegal transfers of executive power" from 1963 to 1977 was linked to a systematic tendency to import Soviet and COMECON arms (Tausch, 1989a, 649). It is worth quoting at length from this study, for Tausch (1989a, 662–663) found that USSR client status allows some respite from large-scale political violence and allows a regime to get cheap weapons on hard currency terms. But some of the root causes of executive instability, of political violence... are among the preconditions that move a regime to turn East, not West (or opt for neutrality, for that matter) for the basic supply of its armed forces.

Another important question concerns regional variability in regime susceptibility to the invocation of SOEs. Figure 3 zeroes in geospatially on the post-Cold War

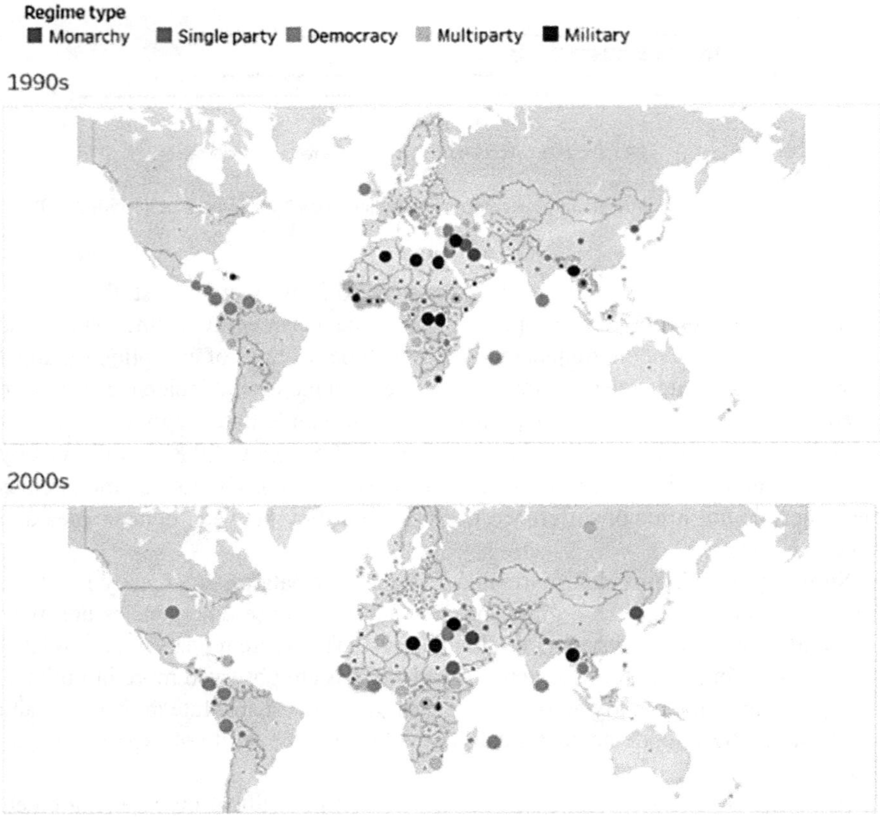

Fig. 3 Spatial and temporal distribution of states of emergency from 1990 to 2009. Note: The dots are scaled according to instances of states of emergency

period. The top panel in the figure reveals that African countries experiencing states of emergency in the 1990s were primarily military regimes. This is important because these regimes, in increasing military expenditure relative to GDP, produce "significant detrimental effects on key development indicators (… the Combined Failed States Index), along with the detrimental effects of high military personnel rates on democracy," and civil and political liberties (Tausch, 1989b, 140; Tausch & Heshmati, 2012, 99). However, as shown for the 2000s (in the bottom panel), multiparty autocracies in Africa were busy enacting emergency measures in the first decade of the century.

Even though SOE invocations are time-and-place-specific contingencies, the data provide evidence of the endogenous nature of domestic conflicts experienced by military and multiparty regimes. The top panel in Fig. 4 plots the average annual probability of civil war by regime type, while the bottom panel displays average levels of mass mobilization, again by regime type. Mass mobilization refers to the mobilization of citizens for mass events such as demonstrations, strikes, and sit-ins,

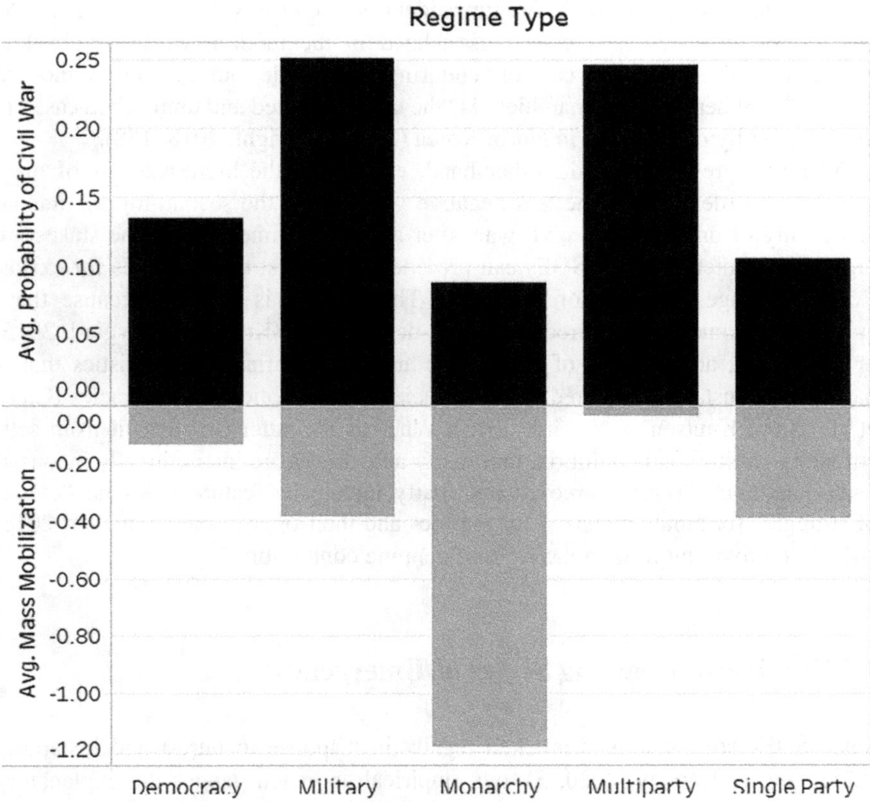

Fig. 4 Domestic conflict by regime type

and the variable can range from −3 to 3, with more negative values indicating less collective action and more positive values more.

It is evident from Fig. 4 that monarchies experience both the lowest levels of mass mobilization (denoted by the more negative values) and the lowest annual probabilities of undergoing a civil war. Monarchical rule in the modern world is either constitutional or absolute. Absolute monarchs remain in the Kingdom of Eswatini, the Nation of Brunei, the Sultanate of Oman, the State of Qatar, and the Kingdom of Saudi Arabia. The Kingdom of Bahrain and the State of Kuwait have representative bodies of some kind, but the monarch retains most of his powers. Likewise, the Kingdom of Jordan, Malaysia, the Kingdom of Morocco, and the United Arab Emirates are constitutional monarchies, but their monarchs still retain more substantial powers than their European counterparts.

Constitutional monarchies are consolidated democracies, whereas absolute monarchies can be regarded as an extreme case of a personalized regime in which power-sharing is minimal or non-existent (Menaldo, 2012). Power is so concentrated in these regimes that in the first type of model we estimate, when interacting regime type and personalism, the interaction term capturing the conditional effect of non-personalism on monarchical regimes identifies no observations in the data. Few polities, in other words, are as personalized in the modern world as absolute monarchies. The only other case of "enduring family rule found . . . in the modern era outside of hereditary monarchies" is "the unprecedented and unmatched case of" the Kim family dictatorship in North Korea (Song & Wright, 2018, 158).

Multiparty regimes, on the other hand, experience the highest levels of mass mobilization (denoted by the less negative values) and the second-highest annual probability of undergoing a civil war, after military regimes. When the stakes are high in such dictatorships, SOEs can provide the greatest opportunities for executives to change the rules of the game. This change is feasible because these autocracies, sometimes referred to as semi-democracies (Knutsen & Nygård, 2015) or anocracies, are a fusion of democratic and authoritarian characteristics that is particularly unstable compared to democracies and consolidated autocracies (Gates et al., 2006; Knutsen & Nygård, 2015). Whereas the latter two benefit from self-enforcing institutional equilibria, multiparty autocracies are institutionally inconsistent (Gates et al., 2006). Moreover, multiparty autocracies feature additional "arenas of struggle" [original emphasis] for regimes and their oppositions (Schedler, 2013, 74) while providing little in the form of genuine contestation.

4 Empirical Modeling States of Emergency

Since SOEs are sometimes invoked legally in response to unrest and instability (Lührmann & Rooney, 2020, 3), our empirical approach gauges the explanatory power of variables frequently used to explain one of their possible outcomes, state repression. From the "standard model" of repression (Keith, 2012, 11), which scholars have refined since Poe and Tate (1994) first introduced it (Richards et al., 2015), we derive the following covariates: civil war, regime type, level of development, economic growth, population size, the size of the armed forces (Albertus & Menaldo, 2012, 152) and personalism (Frantz et al., 2020). In the original model, regime type referred to the presence of political democracy as measured by the Polity project (Keith, 2012, 83).

The level of development is important to consider because "economic impoverishment and political repression are mutually reinforcing." (Solomon & Tausch, 2019, 262). Personalism refers to power accretion by the executive vis-à-vis a ruling party and/or the military. The dynamic this concept conveys differs from the personalist regime category Geddes (1999) originally introduced. As Magaloni et al. (2013) make clear, personalism (the degree to which a regime is tied to one specific individual) is a varying feature of regimes rather than a category in and of itself.

The indicator we use from Geddes et al. (2018) is based on ten "observable indicators of personalist power" (Song & Wright, 2018, 162), such as whether the security apparatus is controlled personally by the leader. See Table 1 in Song and Wright (2018, 163) for definitions of the items used to construct the Personalist

Table 1 Nonlinear panel regression model estimates of states of emergency

	Model [1]	Model [2]	Model [3]
	CRE probit w/time dummies	BTSCS logit w/time dummies	BTSCS logit w/splines
Autocracies (reference: democracies)			
Military($t-1$)	0.243	−0.482	−0.636
Monarchy($t-1$)	−0.417	−2.276***	−1.653**
Multiparty($t-1$)	1.483***	−0.639	−0.073
Single party($t-1$)	0.582	−0.500	−0.235
Personalism($t-1$)	1.359***	1.375*	1.068
Interactions			
Military($t-1$) × personalism($t-1$)	−1.210*	−1.924**	−1.133
Multiparty($t-1$) × personalism($t-1$)	−1.793**	−1.650**	−1.454*
Single party($t-1$) × personalism($t-1$)	−0.747	−2.042**	−1.489
Controls			
Judicial constraints($t-1$)	−2.065***	−0.634	−0.501
Legislative constraints($t-1$)	−1.797***	−0.448	−0.688
Mass mobilization	0.543***	0.476***	0.391***
Civil war	1.459***	0.777***	0.613**
Ethnic fractionalization index	−14.968***	−0.552	−0.448
Log of GDP per capita	−0.159	−0.053	−0.159
Economic growth	−1.023	−0.674	−0.290
Natural resource income	−0.038	0.021	0.020
Political equality	−0.206	0.218**	0.136
Number of coups($t-1$)	−0.054	0.029	0.017
Number of emergencies($t-1$)	−0.017	0.000	−0.003
Regime duration	−0.028***	−0.007	−0.015
Log of population size	−2.310***	−0.247*	−0.179
Size of the military($t-1$)	0.367***	−0.019	0.008
By BTSCS method			
Spell count		−0.043**	−0.673***
Spline 1			−0.016***
Spline 2			0.006***
Spline 3			−0.000
Dependent variables (SOEs)	Instances	Onset	Onset
Total observations (no. of countries)	1890 (68)	2105 (83)	2266 (83)
Clustered means of the covariates	Yes	No	No
Year fixed effects (1952–2006)	Yes	Yes	No
Robust standard errors	No	Yes	Yes
Pseudo R^2		0.167	0.179
rho (Intra-class correlation)	0.859		

Notes: One interaction dropped due to multicollinearity. Standard errors and constant not shown. CRE stands for correlated random effects. BTSCS refers to binary time-series-cross-section data analysis. Cubic splines are used to fit a nonlinear model of emergencies. Two-tailed significant tests reported at $*p < 0.1$; $**p < 0.05$; $***p < 0.01$

Index. As Tausch and Heshmati (2012, xxiii) warned, regimes with a security apparatus that acts as a "state within a state" can engage in widespread violations of human rights. Because the index does not follow a normal distribution and various transformations fail to make it normal, we created a dichotomous variable taking a value of 1 if there is any personalism in the polity and 0 otherwise.

SOEs can also accompany or result from irregular power grabs such as coups and autocoups. Coups are attempts "by the military or other elites within the state apparatus to unseat the sitting head of state using unconstitutional means" (Powell & Thyne, 2011, 252). "A coup attempt is defined as successful if the coup perpetrators seize and hold power for at least seven days" (Powell & Thyne, 2011: 252). From the literature on coups and self-coups (Chap. 3 in Geddes et al., 2018), we thus derive additional predictors: regime duration, mass mobilization, ethnic fractionalization (Svolik, 2015), total resource (petroleum, coal, natural gas, and metals) income per capita, and the distribution of power based on economic (in)equality in society (Houle, 2018).

Regime duration allows us to control for the possibility that the longer a regime has lasted, the more it is likely to survive (Ginsburg & Huq, 2018: 56). Alternatively, regimes have different propensities to last (Przeworski et al., 2000): "average tenure among monarchs [from 1789 to the present for example] is nearly 16 years while average tenure among non-monarchs is nearly 5 years" (Gerring et al., 2021, 609). Also, hybrid regimes "lasted only 9 years on average over the period 1800–2000. In contrast, the average autocracy and democracy endured 21 and 23 years, respectively" (Knutsen & Nygård, 2015).

If coups reflect political instability, they can also create it. This would explain the finding that "coups breed more coups" (Haggard & Kaufman, 2016, 315; Croissant et al., 2017, 99; Svolik, 2012, 136–137). While this might lead us to expect that coups predict autocratization, coups in the recent past "strongly predict democratization, as they indicate key periods of instability and regime weakness" (Miller, 2017, 37). While being unsure of the effect of coups on SOEs, we add to the analysis a count of the number of coup attempts (whether successful or not) per country since 1950. Finally, we add a count of the number of years since 1946 in which a state of emergency was in force in a country (Bjørnskov & Voigt, 2018a, 114). The appendix provides complete information on variable definitions, operationalization, and sources.

5 Accounting for Nonlinearity

To explain the incidence of states of emergency, we adopt a nonlinear model specification which allows us to estimate the effect of regime types while controlling for legislative and judicial constraints on the executive. This yields the following model specification:

$$\text{Probability}(Y_{it}=1|X_{it},C_i)=\Phi(X_{it}\beta+C_i),\ \ i=1,\ldots,N,\ \ t=1,\ldots,T.$$
$$\text{Conditional Distribution } (C_i|\,X_{i1},\ldots,X_{iT})=(C_i|\,\overline{X}_i) \tag{1}$$

where i is a country-level observation and t is the year of each observation (from 1952 to 2006). Y_{it} is a dichotomous dependent variable and X_{it} is a vector of observed explanatory variables. C_i is unobserved country-level heterogeneity. Φ denotes the standard normal cumulative density function. $\overline{X}_i$ is the country-specific mean generated on samples with complete data on all variables (Schunck, 2013, 68). We also lag measures of regime type, personalism, constraints on the executive, and military size by one year to avoid endogeneity bias.

For Eq. 1, we employ a correlated random-effects (CRE) probit model for panel data (Wooldridge, 2011). A key assumption of the traditional random-effects model is that unit-level heterogeneity is uncorrelated with time-varying covariates. More formally, $D(C_i|\,X_{i1},\ldots,X_{iT})=D(C_i)$.

Fixed effects, on the other hand, usually lead to an incidental parameters problem in nonlinear models. "The correlated random-effects model relaxes the assumption of zero correlation between the level 2 error and the level 1 variables" (Schunck, 2013, 67). This flexibility is allowed by introducing the cluster mean of time-varying covariates, which pick up any correlation between these variables and the level 2 (or cluster) error. The CRE model also allows for the inclusion of variables that do not vary within clusters (of which we have none). The quantity of interest to estimate is the average partial effect (or the population average effect) of the deviation scores—or the difference between the cluster mean of a covariate and a particular country-year value. A full battery of yearly dummies on the right-hand side of the regression relaxes the assumption of independence of observations over time within countries.

For the second part of the analysis, we estimate logit models of the onset of emergency spells using the time-series-cross-section with a binary dependent variable (BTSCS) analysis technique (Beck et al., 1998). As Beck et al. show, BTSCS data are group duration data and analysis of such data correctly accounts for the lack of independence units in the same cluster are likely to exhibit. This time dependence can be accounted for in a couple of different ways: using a full set of time dummies like in the previous analysis, or cubic smoothing splines. BTSCS data analysis also requires creating a variable, "spell count," reflecting the number of years by cluster that have elapsed since the last emergency spell onset or since the country entered the study. To meet this requirement, we adopt the logit model specification suggested by Beck et al.:

$$\text{Probability }(Y_{it}=1|X_{it})=1/\left[1+e^{-(X_{it}\beta+K_{t-t0})}\right]$$
$$i=1,\ldots,N,t=1,\ldots,T. \tag{2}$$

where t_0 takes a value of 0 for the year the country first enters the database. Otherwise, t_0 represents the timing of the previous event. t–t_0 is the duration of non-event years between t_0 and t. K_t–t_0 is therefore a count variable that captures the length of sequential zero values before the current observation. In addition, to capture any unobserved country-level heterogeneity, we specify robust standard errors by country. Moreover, BTSCS analysis requires that we have no missing data on the dependent variable, which we proceed to ensure. The number of observations we lose as a result is small, only 90 out of an original 7953.

Part three of the empirical analysis examines emergencies' role in autocratization episodes. Following the Episodes of Regime Transformation (ERT) dataset (Edgell et al., 2020), we operationalize autocratization episodes as "substantial and sustained decreases" in the V-Dem electoral democracy 0–1 index (EDI). The EDI is based on Dahl's (Dahl, 1971) conception of democracy/polyarchy, and an episode requires an initial −0.01 decrease in the EDI "and a total decrease of at least −0.10 throughout the episode."

An autocratization episode ends the final year of a negative change less than or equal to the initial decrease (e.g., −0.01), prior to experiencing an annual increase, cumulative increase, or stasis period. These are defined in the defaults as +0.03, +0.10, and 5 years, respectively (Edgell et al., 2020: 18).

Because the variable asks whether an autocratization episode is ongoing during a particular country-year, it is dichotomous and lends itself to BTSCS modeling. However, a key difference between this and the first two sets of analyses is that political repression is implied in the definition of an autocratization episode. Also, this repression is allegedly invoked in response to events such as mass mobilization, guerrilla warfare, and terrorist attacks. As such, we follow a three-pronged approach to analyzing the effect of political characteristics and states of emergency on autocratization episodes.

First, we remove the variables mass mobilization and civil war from our analysis. Second, because autocratization episodes sometimes unfold over several years, we use current rather than lagged values to estimate the effect of emergencies on these episodes. Thirdly, since regimes vary in how constrained their executives are, we first estimate models of unconditional and conditional (on personalism) regime effects that omit executive constraints (Models 4 and 5). We then estimate unconditional and conditional (on emergencies) models of the effect of legislative and judicial constraints that exclude the variables regime type and personalism (Models 6 and 7).

Our classification of regimes is based on the regime type that exists in a particular country at the end of the year. As Magaloni et al. (2013, 4) point out, "this is the more 'conventional' approach" in datasets of political regimes. To prevent the effect of emergencies from confounding its causes, we thus lag the political characteristics of a country such as its regime type, personalism, the size of its military, and its accumulated number of coups and emergency incidences by one year.

Cross-national analyses of regime change usually control for the spatial and temporal diffusion of regimes, especially democracy. See Houle et al. (2016) for a recent reconsideration of this literature. This is done using the average level of democracy in a country's neighborhood in a particular year or a measure of the number of democratic regimes in a country's neighborhood or the world. Multiparty autocracies have increased in number since the end of the Cold War because of economic leverage exerted by democracies on polities that were once single-party autocracies (Levitsky & Way, 2010; Meng, 2020; Miller, 2017). One benefit of our approach is that it allows these stimuli to affect countries differently and to vary over time in magnitude.

6 Empirical Findings

Table 1 presents the results of our analysis of the incidence (Model 1) and onset (Models 2 and 3) of emergency measures. As expected, personalism accelerates the frequency of states of emergency and the onset of emergency spells. The most revealing part of the results is the effect of regime types on SOEs and how personalism conditions them. Compared to democracies, multiparty regimes experience significantly more SOEs. Once personalized, however, they spend significantly less time under emergencies than democracies.

Turning now to the effect of regime types on the onset of an SOE spell (Models 2 and 3), we find that monarchies, for the reasons noted, exert an unconditional and statistically significant delay on this dependent variable. Our model estimates of the magnitude of this delay vary slightly by specification; in addition, the lagged interaction between a military regime and personalism is significant at the 95 percent confidence level in the model with time dummies (Model 2) but insignificant in the model with cubic splines (Model 3). The variation is indicative of the heterogeneity in the data; specifically, the reality that military regimes were more prone to declare emergencies at the height of the Cold War than since (see Fig. 2 for the time-dependent, regime-specific distribution of SOEs).

Still, in a personalized multiparty regime, we uncover results that are consistent with a model predicting less frequent SOEs and delays in spell onsets. We address this expectation graphically for the marginal effects of multiparty (versus military and single-party) autocracies on SOEs as regime type interacts with personalism. As Fig. 5 indicates, the likelihood of multiparty autocracies experiencing SOEs diminishes (roughly by 21 percent) when they have become personalized. When addressing the onset of SOEs, we find additional evidence (Model 2), although somewhat smaller in magnitude (roughly about 5–9 percent), that personalism significantly delays the probability of SOE declarations in all three types of autocracies (i.e., military, multiparty, and single party) compared to democracies. Controls included in Table 1 show anticipated results. Mass mobilization, civil wars, and the size of the military apparatus all significantly increase emergency incidence as well as accelerate the onset of spells in the case of the first two variables.

Both population size and regime duration decrease this incidence. Scholars expect population size to increase state repression. However, the negative and statistically significant effect of this variable suggests that larger populations might make it harder for governments to achieve their objectives using SOEs. Finally (and as expected), the relationship between the time elapsed since the beginning of the

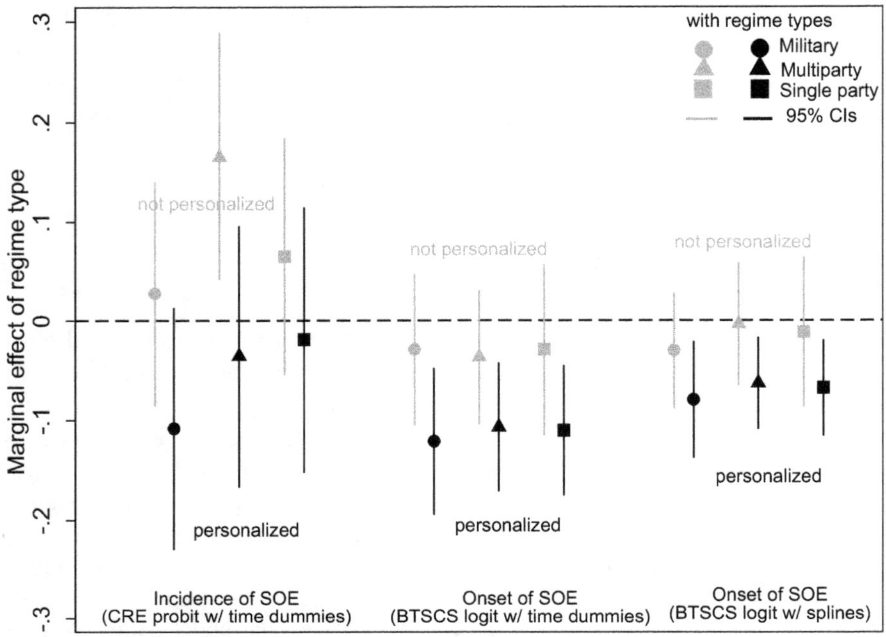

Fig. 5 Marginal effects of regime types interacting with personalism on the probability of SOEs. Note: For marginal effect calculations, we used the Stata margins command, leaving all other independent variable values as they are

last spell (as indicated by the spell count variable) and the outcome variable is negative and statistically significant. Altogether, the results reveal that most covariates responsible for the incidence of emergencies also explain the time that elapses before a new emergency spell begins.

Bjørnskov and Voigt (2018a, 111) claim that executives "react to constitutional incentives by calling an SOE when . . . they do not face a formally independent judiciary or other veto players." We also find that lagged judicial and legislative constraints significantly lessen the frequency of SOEs (Model 1) although they do not significantly delay their declaration (Models 2 and 3). Moreover, these effects are robust even after controlling for the type of regime in the polity, personalism, and their interaction in the year before the declaration of an SOE.

As Table 2 indicates, we do not find that vulnerability to autocratization is significantly higher among authoritarian regimes compared to their democratic counterparts, nor do we find that prior coups raise the probability of autocratization. The nonlinear estimates of the interaction terms in Models 4 and 5 show however that personalized multiparty dictatorships are systematically associated with substantial and sustained autocratization. Hybrid regimes, in other words, are significantly likely to become more autocratic when emergencies are invoked and the regime becomes more personalistic. Strong personalism characterizes the dictator's ability to remove elites 'from the positions that grant them power and authority'

Table 2 Nonlinear panel regression model estimates of autocratization episodes

Variables	Model [4] BTSCS logit w/time dummies	Model [5] BTSCS logit w/splines	Model [6] BTSCS logit w/time dummies	Model [7] BTSCS logit w/splines
Autocracies (reference: democracies)				
Military regime(*t*−1)	−0.716	−0.475		
Monarchy(*t*−1)	0.646	0.951		
Multiparty regime(*t*−1)	−1.515	−0.703		
Single-party regime(*t*−1)	0.180	0.725		
Personalism(*t*−1)	−0.660	−0.264		
States of emergency			0.711**	0.316
Judicial constraints(*t*−1)			1.590***	0.982*
Legislative constraints(*t*−1)			0.062	0.605
Interactions				
Military regime(*t*−1) × personalism(*t*−1)	1.102	0.901		
Multiparty regime(*t*−1) × personalism (*t*−1)	2.574***	2.149***		
Emergency × judicial constraints(*t*−1)			−1.104*	−0.583
Emergency × legislative constraints(*t*−1)			1.410**	1.550***
Controls				
Ethnic fractionalization index	−0.206	−0.307	0.065	0.106
Log of GDP per capita	−0.260	−0.223	−0.494***	−0.514***
Economic growth	−1.327	−1.055	−1.244*	−0.424
Natural resource income	−0.036	−0.030	−0.020	−0.004
Political equality	0.301***	0.239**	−0.137*	−0.098
Number of coups(*t*−1)	−0.074**	−0.069**	−0.010	−0.012
Number of emergencies(*t*−1)	0.028*	0.008	−0.005	−0.011
Regime duration	0.008	−0.009	−0.008	−0.017***
Log of population size	−0.020	−0.018	−0.017	−0.029
Size of the military(*t*−1)	0.007	−0.014	0.112**	0.080
Dependent variable	Autocratization			
Using BTSCS method				
Spell count	−0.190***	−0.866***	−0.145**	−0.896***
Spline 1		−0.008***		−0.008***
Spline 2		0.004***		0.004***
Spline 3		−0.001***		−0.001***
Total observations (No. of countries)	2741 (106)	3274 (109)	5662 (143)	5706 (143)
Year fixed effects (1952–2006)	Yes	No	Yes	No
Wald chi2	163.63***	311.61***	398.11***	686.47***
Pseudo R^2	0.274	0.365	0.282	0.373

Notes: Two interactions dropped due to multicollinearity. Standard errors and constant are not shown. BTSCS refers to binary time-series-cross-section data analysis. Cubic splines are used to fit a nonlinear model of emergencies. Two-tailed significant tests reported at $*p < 0.1$; $**p < 0.05$; $***p < 0.01$

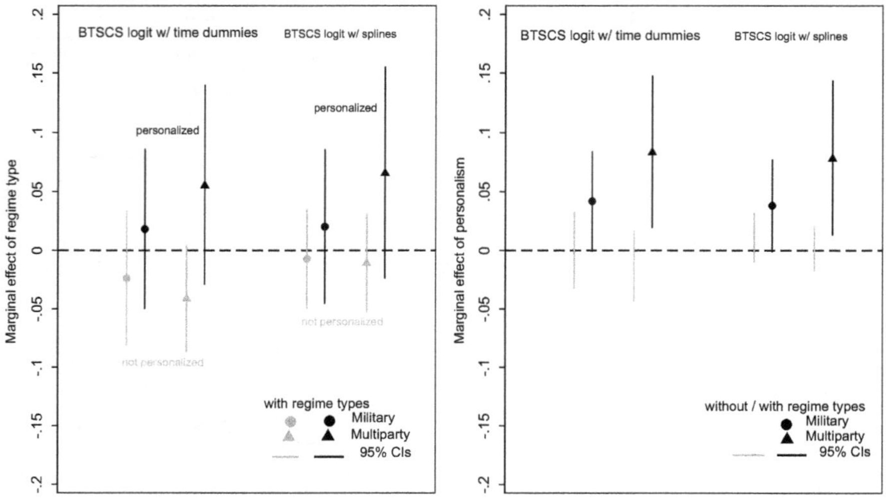

Fig. 6 Interactive effects of regime types and personalism on the probability of autocratization. (**a**) Marginal effects of regime types interacting with personalism on the probability of autocratization. (**b**) Marginal effects of personalism interacting with regime types on the probability of autocratization. Note: For marginal effect calculations, we used the Stata margins command, leaving all other independent variable values as they are

and thus hold greater power over these elites (Goldring & Matthews, 2021, 4). Highlighted in models 6 and 7 are the effects of emergencies on autocratization episodes. As expected, emergencies significantly accelerate the onset of these episodes (Model 6), while the more time passes since the last episode, the more successfully a country can avoid a new one (Models 4–7).

As a robustness check, we look at the effect of personalism on autocratization conditional on regime types. See Fig. 6a. This is also done if interactions are asymmetric (Berry et al., 2012) as in Fig. 6b. Therein, we show how the autocratizing effect of personalism varies with regime type. In this subfigure, we find an increasing marginal effect on the probability of autocratization that is significantly different from zero. This conditional effect ranges roughly from 3 to 9 percent whereas the logit model with year dummies predicts a slightly larger increase.

7 Sensitivity Analyses

To check the robustness of our findings to sampling issues, we ran a range of sensitivity tests addressing overall model performance and adding relevant contextual variables.

Sampling: Cross-national data over long periods are likely to include observations that are not missing at random; this creates systematic bias. Similarly, personalism is a power accretion dynamic taking place at the discretion of a regime's key stakeholders that is often hidden from public scrutiny. Therefore, it comes as no surprise that 42.7 percent of SOE cases from the V-Dem database do not have values for our personalism indicator.

To adequately address the risk of sample selection bias, we conducted panel Jackknife analyses by withdrawing one country from the pooled sample and replacing it with another country iteratively. See Fig. 7. This methodology helps to identify which observation(s) (if any) drive the statistical association between personalism and states of emergency. Based on our analysis, the most influential country-year observations are the reigns of leaders who came to power through a military coup (e.g., Park Chung-hee and Chun Doo-hwan in South Korea, Ibrahim Babangida in Nigeria, and Pervez Musharraf in Pakistan). These individuals leaped at the opportunity to consolidate their power by declaring an SOE, especially early in their tenure when the rules of the game were highly uncertain (Frantz, 2018). Despite this sampling issue, we find that the dampening effect of personalized multiparty regimes on SOEs persists.

Omitted variables: To minimize concerns with omitted variable bias, we include a measure for civil war duration as a relevant time-varying variable in our benchmark model specification (Model 1, Table 1). As Brambor et al. (2006) recommend, we also added more interaction terms (e.g., for joint effects of civil war duration and regime type) to a nonlinear function of SOEs. Note that we used the duration of civil wars instead of their incidence as a contextual variable. This is because empirically, the former can better trace various conflict stages such as emergence, escalation, "hurting" stalemate, negotiation, and dispute settlement (Zartman, 1989). We find support for the idea that civil war duration plays a contextual role in determining the frequency of SOEs over conflict stages. See Table 3 for the full report and Fig. 8 for

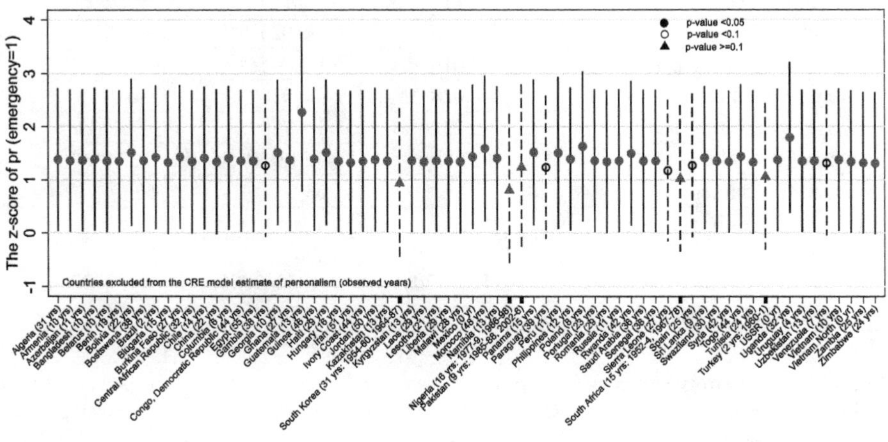

Fig. 7 Jackknife analyses of personalism effects (from Model 1, Table 1)

Table 3 Robustness checking to alternative model specifications (civil war duration)

	Model [8]	Model [9]	Model [10]
Dependent variable: Instances of SOEs	CRE probit w/time dummies	CRE probit w/time dummies	CRE probit w/time dummies
Autocracies (reference: Democracy)			
Military ($t-1$)	0.338	−0.362	0.127
Monarchy ($t-1$)	−0.498	−0.540	−0.939
Multiparty ($t-1$)	1.487**	0.263	1.103*
Single party ($t-1$)	0.613	0.352	0.323
Personalism ($t-1$)	1.423**	0.276	1.070
Civil war instance	1.337***	0.915***	0.920***
Civil war duration	0.024	−0.040	−0.040
Interactions			
Military ($t-1$) × personalism ($t-1$)	−1.273*		−1.057
Multiparty ($t-1$) × personalism ($t-1$)	−1.821**		−1.454*
Single-party ($t-1$) × personalism ($t-1$)	−0.805		−0.387
Military ($t-1$) × civil war duration		0.440***	0.450***
Multiparty ($t-1$) × civil war duration		0.258***	0.250***
Controls			
Legislative constraints ($t-1$)	−1.783***	−0.981*	−1.100**
Judicial constraints ($t-1$)	−2.053***	−2.157***	−1.923***
Mass mobilization	0.543***	0.534***	0.529***
Ethnic fractionalization index	−14.705***	−16.487***	−17.423***
Log of GDP per capita	−0.162	−0.184	−0.188
Economic growth	−1.029	−1.118	−1.180
Natural resource income	−0.039	−0.037	−0.026
Political equality	−0.212*	−0.184	−0.195
Number of coups ($t-1$)	−0.058	−0.023	−0.024
Number of emergencies ($t-1$)	−0.016	−0.044**	−0.049**
Regime duration	−0.028***	−0.018**	−0.019**
Log of population size	−2.228***	−1.789**	−2.113***
Size of the military ($t-1$)	0.359***	0.313***	0.334***
Number of countries	68	68	68
Clustered means of the covariates	Yes	Yes	Yes
Year fixed effects	Yes (1952–2006)	Yes (1952–2006)	Yes (1952–2006)
rho (Intra-class correlation)	0.857	0.856	0.859
Total observations	1890	1890	1890

Notes: Standard errors and constants are not shown. The reference category for personalism is regimes that are not personalistic. CRE stands for correlated random effects. Two-tailed significant tests reported at $*p < 0.1$; $**p < 0.05$; $***p < 0.01$

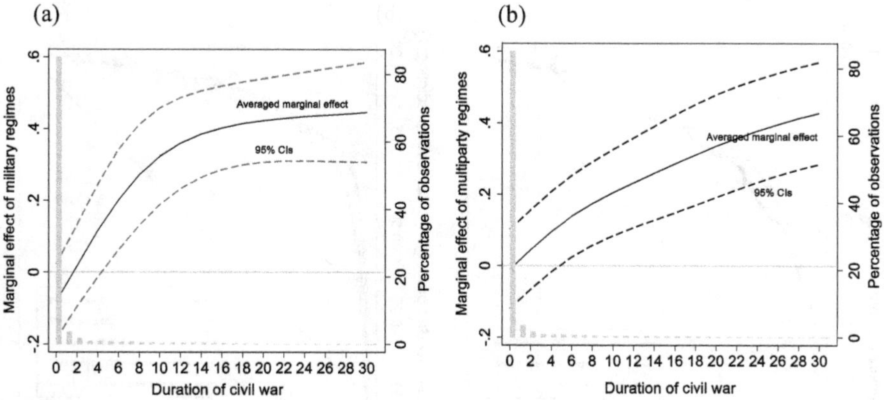

Fig. 8 Interactive effects of autocracies and civil war duration on the probability of emergency declarations. (**a**) Marginal effect of military regimes interacting with civil war duration on the probability of emergency declarations. (**b**) Marginal effect of multiparty regimes interacting with civil war duration on the probability of emergency declarations. Note: For marginal effect calculations, we used the Stata margins command, leaving all other independent variable values (in Model 3, Table 3) as they are

the marginal effect plots that capture this contextual dependence. Controlling for this additional contextual effect does not alter our key finding substantively.

Model performance: To diagnose logit model performance, we plot ROC (Receiver Operating Characteristic) curves. Post-estimates from the emergency onset models (from Table 1) and the autocratization models (from Table 2) show large percentages of binary outcome observations that are correctly predicted for a positive outcome (i.e., the event occurred): Model 2 (94.16%), Model 3 (94.57%), Model 4 (94.71%), Model 5 (96.21%), Model 6 (94.90%), and Model 7 (95.55%). To obtain these quantities of interest, we used the Stata estat classification command. This command works with estimation results from BTSCS logit (but not those from CRE probit).

These high success rates are based on the method of using 0.5 as an arbitrary cut-point for predicted probabilities. In so doing, we could overstate the probability that a given observation is a positive outcome based on the values of the explanatory variables. In other words, our arbitrary choice of a cut-point can be biased so that any observation with a fitted probability above the cut-point is classified as positive. In contrast, any observation below the cut-point is classified as negative. For instance, based on the arbitrary cut-point of 0.5, we treat observations with a fitted probability of 0.51 in the same way as observations with a fitted probability of 0.99.

To check how the model distinguishes between positive and negative outcomes as we move this arbitrary cut-point from 0 to 1, we look at the area under the parametric binomial ROC curves. We used the Stata roccomp command (available for BTSCS logit) to perform this simulation. The idea is that the greater the area under the ROC curve, the better the model performs at correctly classifying outcomes. Figure 9a and b compare the performance of multiple models. We find that

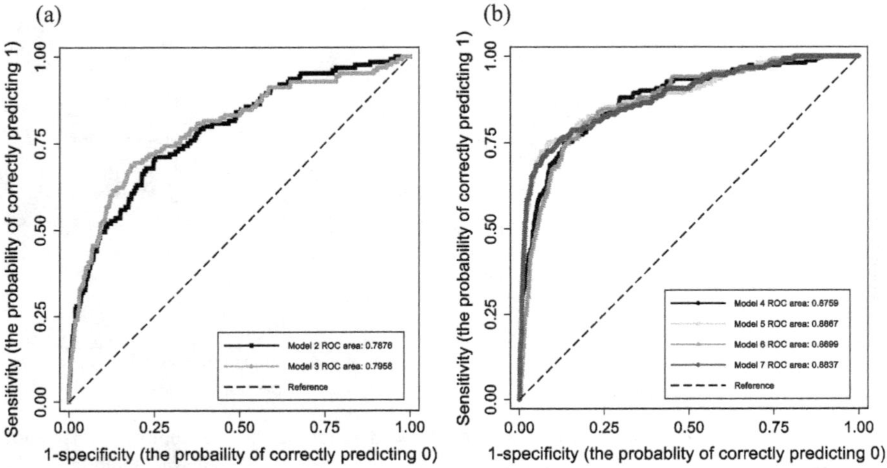

Fig. 9 Diagnostic abilities of a binary classifier system using the receiver operating characteristic (ROC) curve. (**a**) Models of emergency onset (yes = 1, no = 0) from Table 1. (**b**) Models of autocratization (yes = 1, no = 0) from Table 1

models with splines where the conditional effect of multiparty regimes subject to personalism is considered (Model 3 from Table 1; Model 5 from Table 2) outperform models with time dummies (Models 2 and 4).

8 Concluding Observations

States of emergency are common political events, but little empirical work has been done to understand their causes, relationship to regime types, and other institutional factors that account for their frequency. Inspired by the work of Arno Tausch, we sought to fill this gap, first with a descriptive analysis of when and where emergencies occur. We then developed a nonlinear empirical model for heterogeneity in the extent to which executives declare emergencies to change the balance of power within a regime.

Emergencies affect both democracies and nondemocracies alike, but we find that monarchies experience significantly less and multiparty dictatorships significantly more emergencies than other regime types, including democracies. However, less frequent SOEs result when powerful executives lead multiparty autocracies. Personalized multiparty regimes also show longer spells in between SOEs while they trigger more autocratization episodes.

The reader could be forgiven for thinking that absolute monarchies come out looking very well in our analysis. We wish to caveat our results with the observation that where Muslim rulers continue to rely on religion for their legitimacy.

> Governance ... reflects the narrow interests of a political leadership increasingly out of touch with reality, a conservative religious establishment fearful of change and military elites who view the common citizen as an enemy. Because economic interests are not incorporated at this bargaining table, few pro-growth policies are adopted, revenues of the state shrink and with it state capacity to deliver basic services to its population who grow increasingly restive. This leads to a vicious circle of ever greater repression on the part of the security services who increasingly come to have a greater say in governance. Ultimately, they come to control the state as in the case of Egypt where Al-Sisi seamlessly moved from the military rank of "General" to the "civilian rank" of President. Ultimately, the status of failed states is the natural outcome of these polities as represented by Yemen (Solomon & Tausch, 2019, 56).

While we believe Solomon and Tausch are right about the non-monarchies in the MENA, recent work (Menaldo, 2012) and our data in Fig. 4 indicate that the argument is more applicable to the republics than the monarchies of the region. As we saw, absolute monarchies experience very little mass mobilization.

We limited our nonlinear model analysis to emergencies declared in response to overt domestic conflict. We did this because we believe that although emergencies are sometimes declared in response to events such as natural disasters, there is no inherent link between SOEs declared following such disasters and conflict behavior that creates new political opportunities (e.g., Omelicheva, 2011). We thus believe that SOEs declared in response to events such as rioting and terrorist attacks are more consequential. SOE observations might have been subject to data truncation that results in some loss of information (roughly 4% of our country-year observations). The emergencies that remain, however, are the ones most bearing on questions of conflict and order.

The scope of personalist variation is still narrow (and could be biased) due to empirical limitations. Future research should tend to this question as well as explore why judicial and legislative constraints significantly lessen the incidence of SOEs but do not delay their declaration. One possibility is that checks on the executive, rather than behaving uniformly across regimes, vary in their workings across or within them. Bjørnskov and Voigt (2018a, 119) provide evidence that this is the case since they expect parliamentary democracies to call for more SOEs than presidential ones, but their analysis uncovers the reverse. Considering this finding, we did not differentiate checks and balances by regime type. Presidential systems, however, typically have more veto players than parliamentary ones and as a result, should experience more SOEs according to both their logic and ours.

Variable Definitions, Summary Statistics, and Sources

Variables	Definitions	Min	Max	Mean	Std. Dev.	Source
State of emergency						
Instances	Declaring a national state of emergency due to a terrorist attack, an armed conflict/war (domestically), and/or mass protest/popular uprising. 1 if a state of emergency is declared, 0 otherwise. Transformed to a dichotomous value.	0	1	0.32	0.47	V-Dem Dataset (Ver. 10) (Coppedge et al., 2020). The Episodes of Regime Transformation dataset (Edgell et al., 2020).
Onsets		0	1	0.05	0.22	
Autocratization Episode						
Onsets	The onset of a gradual decline in democratic qualities. Following Edgell et al. (2020), the onset of autocratization episodes was coded as 1 if the electoral democracy index decreased by 1% and its total decrease was at least 0.1 throughout the episode. Coded as 0 otherwise.	0	1	0.05	0.22	
Authoritarian regime types						
Military	Mutually exclusive dichotomous coding was applied for a given country-year observation. Democracy is used as the reference category.	0	1	0.18	0.38	The Autocracies of the World dataset (Magaloni et al., 2013)
Monarchy		0	1	0.07	0.26	
Multiparty		0	1	0.18	0.38	
Single Party		0	1	0.19	0.39	
Personalism	Power accretion in relation to a ruling party and/or the military rather than other branches of government. Transformed to a dichotomous value. 1 if any personalism, 0 otherwise.	0	1	0.84	0.37	Geddes et al. (2018)

(continued)

Variables	Definitions	Min	Max	Mean	Std. Dev.	Source
Judicial constraints	An index of judicial constraints on the executive, with a larger value indicating the executive is more respectful of the constitution, more in compliance with court rulings, and there are more de facto conditions for judicial independence.	0.01	0.99	0.51	0.31	V-Dem Dataset (Ver. 10) (Coppedge et al., 2020)
Legislative constraints	An index of legislative constraints on the executive, with a larger value indicating the legislature is capable of questioning, investigating, and exercising checks on the executive.	0.00	0.99	0.43	0.35	
Mass mobilization	Continuous measure derived from ordinal (5-level) measure of the mobilization of citizens for mass events including demonstrations, strikes, and sit-ins.	–3.20	3.32	−0.30	1.29	V-Dem Dataset (Ver. 10) (Coppedge et al., 2020)
Ethnic fractionalization index	Ethnic fractionalization index from the Historical Index of Ethnic Fractionalization Dataset (HIEF)	0.00	0.89	0.42	0.28	Drazanova (2019)
GDP per capita	GDP per capita logged base 10.	4.90	11.65	843	1.13	V-Dem Dataset (Ver. 10) (Coppedge et al., 2020)
Economic growth	GDP per capita growth rate	–0.98	3.67	0.03	0.11	V-Dem Dataset (Ver. 10) (Coppedge et al., 2020)
Natural resource income	The real per capita value of a country's petroleum, coal, natural gas, and metal production, logged.	–4.61	11.11	2.53	4.04	V-Dem Dataset (Ver. 10) (Coppedge et al., 2020)
Political equality	A continuous version of a 5-level measure of power distributed by socioeconomic position: the degree with which wealth and income are translated into political power. Converted to a logarithmic scale.	–3.17	3.35	0.44	1.26	UCDP/PRIO Armed Conflict Dataset Version 19.1 (Gleditsch et al., 2002; Pettersson, 2019; Pettersson et al., 2019)

(continued)

Variables	Definitions	Min	Max	Mean	Std. Dev.	Source
Number of coups	The cumulative number of successful and unsuccessful coups.	0	27	2.47	4.11	
Number of emergencies	The cumulative number of years under a state of emergency.	0	65	4.02	8.58	
Civil war	Internal or internationalized internal war	0	1	0.16	0.37	
Regime duration	The number of years the regime has been in power, up to and including the observation year	1	269	4.03	8.58	Geddes et al., 2014
Population size	Population in millions, logged	–1.94	7.22	2.31	1.47	Penn World Table version 10.0 (Feenstra et al., 2015)
Size of the military	The natural logarithmic of military personnel in the thousands	–4.61	8.67	3.61	2.17	Correlates of War National Material Capabilities Dataset (Version 5.0) (Singer et al., 1972; Singer, 1987)

References

Albertus, M., & Menaldo, V. (2012). Coercive capacity and the prospects for democratization. *Comparative Politics, 44*(2), 151–169.

Barany, Z. (2016). *How armies respond to revolutions and why*. Princeton University Press.

Beck, N., Katz, J. N., & Tucker, R. (1998). Taking time seriously: Time-series-cross-section analysis with a binary dependent variable. *American Journal of Political Science, 42*(4), 1260–1288.

Berry, W. D., Golder, M., & Milton, D. (2012). Improving tests of theories positing interaction. *Journal of Politics, 74*(3), 653–671.

Bjørnskov, C., & Voigt, S. (2018a). Why do governments call a state of emergency? On the determinants of using emergency constitutions. *European Journal of Political Economy, 54*, 110–123.

Bjørnskov, C., & Voigt, S. (2018b). The architecture of emergency constitutions. *International Journal of Constitutional Law, 16*(1), 101–127.

Bjørnskov, C., & Voigt, S. (2020). When does terror induce a state of emergency? And what are the effects? *Journal of Conflict Resolution, 64*(4), 579–613.

Brambor, T., Clark, W. R., & Golder, M. (2006). Understanding interaction models: Improving empirical analyses. *Political Analysis, 14*(1), 63–82.

Chang, P. Y. (2015). *Protest dialectics: State repression and South Korea's democracy movement, 1970–1979*. Stanford University Press.

Coppedge, M., Gerring, J., Knutsen, C. H., Lindberg, S. I., Teorell, J., Altman, D., Bernhard, M., Fish, M. S., Glynn, A., Hicken, A., Luhrmann, A., Marquardt, K L, McMann, K., Paxton, P., Pemstein, D., Seim, B., Sigman, R., Skaaning, S.-E., Staton, J., Wilson, S., Cornell, A., Alizada, N., Gastaldi, L., Gjerløw, H., Hindle, G., Ilchenko, N., Maxwell, L., Mechkova, V., Medzihorsky, J., von Römer, J., Sundström, A., Tzelgov, E., Wang, Y., Wig, T., & Ziblatt, D. (2020). *V-Dem [country–year/country–date] dataset v10. varieties of democracy* (V-Dem) project. https://doi.org/10.23696/vdemds20

Croissant, A., Eschenauer, T., & Kamerling, J. (2017). The military and autocratic regimes: Roles of the armed forces. In J. Gerschewski & C. H. Stefes (Eds.), *Crisis in autocratic regimes* (pp. 89–110). Lynne Rienner Publishers.

Dahl, R. A. (1971). *Polyarchy: Participation and opposition.* Yale University Press.

Domínguez, J. I. (2011). The perfect dictatorship? South Korea versus Argentina, Brazil, Chile, and Mexico. In K. Byung-Kook & F. V. Ezra (Eds.), *The Park Chung Hee era: The transformation of South Korea* (pp. 573–602). Harvard University Press.

Drazanova, L. (2019). *Historical index of ethnic fractionalization dataset* (HIEF). https://doi.org/10.7910/DVN/4JQRCL.

Edgell, A. B., Maerz, S. F., Maxwell, L., Morgan, R., Medzihorsky, J., Wilson, M. C., Boese, V., Hellmeier, S., Lachapelle, J., Lindenfors, P., Lührmann, A., & Lindberg, S. I. (2020). *Episodes of regime transformation dataset (v2.0) codebook.* Retrieved August 4, 2022, from https://www.v-dem.net/media/filer_public/73/4a/734a8cb3-5068-4fb9-9318-a99556d978ce/ert_codebook.pdf

Feenstra, R. C., Inklaar, R., & Timmer, M. P. (2015). The next generation of the Penn world table. *American Economic Review, 105*(10), 3150–3182.

Frantz, E. (2018). *Authoritarianism: What everyone needs to know.* Oxford University Press.

Frantz, E., Kendall-Taylor, A., Wright, J., & Xu, X. (2020). Personalization of power and repression in dictatorships. *Journal of Politics, 82*(1), 372–377.

Gates, S., Hegre, H., Jones, M. P., & Strand, H. (2006). Institutional inconsistency and political instability: Polity duration, 1800–2000. *American Journal of Political Science, 50*(4), 893–908.

Geddes, B. (1999). What do we know about democratization after twenty years? *Annual Review of Political Science, 2*, 115–144.

Geddes, B., Joseph, W., & Erica, F. (2014). Autocratic breakdown and regime transitions: A new data set. *Perspectives on Politics, 12*(2), 313–331.

Geddes, B., Wright, J., & Frantz, E. (2018). *How dictatorships work: Power, personalization, and collapse.* Cambridge University Press.

Gerring, J., Wig, T., Veenendaal, W., Weitzel, D., Teorell, J., & Kikuta, K. (2021). Why monarchy? The rise and demise of a regime type. *Comparative Political Studies, 54*(3–4), 585–622.

Ginsburg, T., & Huq, A. Z. (2018). *How to save a constitutional democracy.* University of Chicago Press.

Gleditsch, N. P., Wallensteen, P., Eriksson, M., Sollenberg, M., & Strand, H. (2002). Armed conflict 1946–2001: A new dataset. *Journal of Peace Research, 39*(5), 615–637.

Goldring, E., & Matthews, A. S. (2021). To purge or not to purge? An individual-level quantitative analysis of elite purges in dictatorships. *British Journal of Political Science, 23*, 230–254. https://doi.org/10.1017/S007123421000569

Greitens, S. C. (2016). *Dictators and their secret police: Coercive institutions and state violence.* Cambridge University Press.

Hafner-Burton, E. M., Helfer, L. R., & Fariss, C. J. (2011). Emergency and escape: Explaining derogations from human rights treaties. *International Organization., 65*(4), 673–707.

Haggard, S., & Kaufman, R. R. (2016). *Dictators and democrats: Masses, elites, and regime change.* Princeton University Press.

Hilbink, L. (2007). *Judges beyond politics in democracy and dictatorship: Lessons from Chile.* Cambridge University Press.

Honig, B. (2009). *Emergency politics: Paradox, law, democracy.* Princeton University Press.

Houle, C. (2018). Does economic inequality breed political inequality? *Democratization, 25*(8), 1500–1518.
Houle, C., Kayser, M. A., & Xiang, J. (2016). Diffusion or confusion? Clustered shocks and the conditional diffusion of democracy. *International Organization., 70*, 687–726.
Keith, L. C. (2012). *Political repression: Courts and the law*. The University of Pennsylvania Press.
Keith, L. C., & Poe, S. C. (2004). Are constitutional state of emergency clauses effective? An empirical exploration. *Human Rights Quarterly, 26*(4), 1071–1097.
Knutsen, C. H., & Nygård, H. M. (2015). Institutional characteristics and regime survival: Why are semi-democracies less durable than autocracies and democracies? *American Journal of Political Science, 59*(3), 656–670.
Levitsky, S., & Way, L. A. (2010). *Competitive authoritarianism: Hybrid regimes after the Cold War*. Cambridge University Press.
Lührmann, A., & Rooney, B. (2020). Autocratization by decree: States of emergency and democratic decline. *Comparative Politics, 53*(4), 617–649.
Lührmann, A., & Lindberg, S. I. (2019). A third wave of autocratization is here: What is new about it? *Democratization, 26*(7), 1095–1113.
Magaloni, B. (2008). Credible power-sharing and the longevity of authoritarian rule. *Comparative Political Studies, 41*(4/5), 715–741.
Magaloni, B., Chu, J., & Min, E. (2013). *Autocracies of the world, 1950–2012 (Version 1.0) dataset.* Stanford University. Retrieved August 4, 2022, from https://cddrl.fsi.stanford.edu/research/autocracies_of_the_world_dataset.
Menaldo, V. (2012). The Middle East and North Africa's resilient monarchs. *Journal of Politics, 74*(3), 707–722.
Meng, A. (2020). *Constraining dictatorship: From institutionalized rule to institutionalized regimes*. Cambridge University Press.
Miller, M. K. (2017). The strategic origins of electoral authoritarianism. *British Journal of Political Science, 50*(1), 17–44.
Omelicheva, M. Y. (2011). Natural disasters: Triggers of political instability? *International Interactions., 37*(4), 441–465.
Pepinsky, T. (2014). The institutional turn in comparative authoritarianism. *British Journal of Political Science, 44*(3), 631–653.
Pérez-Liñán, A. (2007). *Presidential impeachment and the new political instability in Latin America*. Cambridge University Press.
Pettersson, T. (2019). *UCDP/PRIO Armed conflict dataset codebook v 19.1*. Retrieved August 4, 2022, from https://ucdp.uu.se/downloads.
Pettersson, T., Högbladh, S., & Öberg, M. (2019). Organized violence, 1989–2018 and peace agreements. *Journal of Peace Research, 56*(4), 589–603.
Poe, S. C., & Tate, N. (1994). Repression of human rights to personal integrity in the 1980s: A global analysis. *American Political Science Review, 88*(4), 853–872.
Policzer, P. (2009). *The rise and fall of repression in Chile*. The University of Notre Dame Press.
Powell, J. M., & Thyne, C. L. (2011). Global instances of coups from 1950 to 2010: A new dataset. *Journal of Peace Research, 48*(2), 249–259.
Przeworski, A., Alvarez, M. E., Cheibub, J. A., & Limongi, F. (2000). *Democracy and development: Political institutions and well-being in the world, 1950–1990*. Cambridge University Press.
Richards, D. L., & Clay, K. C. (2012). An umbrella with holes: Respect for non-derogable human rights during declared states of emergency, 1996–2004. *Human Rights Review, 13*, 443–471.
Richards, D. L., Webb, A., & Clay, K. C. (2015). Respect for physical-integrity rights in the twenty-first century: Evaluating Poe and Tate's model 20 years later. *Journal of Human Rights, 14*(3), 291–311.
Schedler, A. (2013). *The politics of uncertainty: Sustaining and subverting electoral authoritarianism*. Oxford University Press.

Schunck, R. (2013). Within and between estimates in random-effects models: Advantages and drawbacks of correlated random effects and hybrid models. *The Stata Journal, 13*(1), 65–76.
Sheeran, S. P. (2013). Reconceptualizing states of emergency under international human rights law: Theory, legal doctrine, and politics. *Michigan Journal of International Law, 34*(3), 491–557.
Singer, J. D. (1987). Reconstructing the correlates of war dataset on material capabilities of states, 1816–1985. *International Interactions, 14*, 115–132.
Singer, J. D., Bremer, S., & Stuckey, J. (1972). Capability distribution, uncertainty, and major power war, 1820–1965. In B. M. Russett (Ed.), *Peace, war, and numbers* (pp. 19–48). Sage.
Solomon, H., & Tausch, A. (2019). *Islamism, crisis and democratization: Implications of the world values survey for the Muslim world*. Springer.
Song, W., & Wright, J. (2018). The North Korean autocracy in comparative perspective. *Journal of East Asian Studies, 18*(2), 157–180.
Svolik, M. (2012). *The Politics of authoritarian rule*. Cambridge University Press.
Svolik, M. (2015). Which democracies will last? Coups, incumbent takeovers, and the dynamic of democratic consolidation. *British Journal of Political Science, 45*(4), 715–738.
Tausch, A. (1989a). Armas socialistas, subdesarrollo y violencia estructural en el tercer mundo. *Revista Internacional de Sociología., 47*(4), 583–716.
Tausch, A. (1989b). Stable Third World democracy and the European model: A quantitative essay. In Z. Bablewski & B. Hettne (Eds.), *Crisis in development*. United Nations University.
Tausch, A., & Heshmati, A. (2012). *Globalization, the human condition and sustainable development in the twenty-first century: Cross-national perspectives and European implications*. Anthem Press.
Tausch, A., & Prager, F. (1993). *Towards a socio-liberal theory of world development*. Palgrave Macmillan.
Wooldridge, J. M. (2011). A simple method for estimating unconditional heterogeneity distributions in correlated random effects models. *Economics Letters, 113*, 12–15.
Zartman, I. W. (1989). *Ripe for resolution: Conflict and intervention in Africa*. Oxford University Press.

Rabin and the Intelligence Services on the Eve of the Oslo Accords

Jacques Neriah

Abstract This essay stresses that intelligence is at the heart of decision-making in all fields, especially when it is a matter of life and death, the fate of a diplomatic initiative, a decision to start or end a war. In a similar approach, Arno Tausch wrote that during his years as a labor attaché in Warsaw from 1992 to 1999 and in the ministry for social affairs from 1999 to 2016, he never had access to classified intelligence of any source but what he did to the best of his knowledge was to provide his superiors with good and sound social scientific information which revealed to the decision makers many important points. These assessments, according to Arno Tausch, were linked with assessments what the most important think tanks had to say on a given subject, for example, the migration potential of the countries of Eastern Europe to the old member countries of the European Union at the time of European Union enlargement.

Intelligence is at the heart of decision-making in all fields, especially when it is a matter of life and death, the fate of a diplomatic initiative, and a decision to start or end a war. Knowing your enemy, your rival, your competitor, your friend, and your business partner cannot be realized without prior knowledge of intentions, qualities, character, habits, routine, psychology, and health.

Intelligence is in fact at the essence of life and survival. It is the result of treating and analyzing raw information. It is the human conclusion drawn from information gathered on a subject, an object, a topic and therefore it is open to misinformation, to mistakes in judgment, in analysis. Israel's modern history is full of examples about the use of intelligence and about the faulty interpretations of raw information due to conceptual rigidity and prejudice. The blatant example is the October 1973 war in which the analysts failed to interpret the early warning signs collected by the different spy agencies who warned of a coming war.

In a similar approach, Professor Arno Tausch wrote that during his years as a labor attaché in Warsaw from 1992 to 1999 and in the ministry for social affairs from

J. Neriah (✉)
Jerusalem Center for Public Affairs, Jerusalem, Israel

G. Segell (ed.), *Development, Globalization, Global Values, and Security*,
https://doi.org/10.1007/978-3-031-24513-8_14

1999 to 2016, he never had access to classified intelligence of any source but what he did to the best of his knowledge was to provide his superiors with good and sound social scientific information which revealed to the decision makers many important points. These assessments, according to Professor Tausch, were linked with assessments what the most important think tanks had to say on a given subject, for example, the migration potential of the countries of Eastern Europe to the old member countries of the European Union at the time of European Union enlargement. The UK had opened its borders at the time of European Union enlargement to the labor markets from eastern Europe with the influx of millions of workers from the region to the UK in the following years, while Austria kept its labor market tight for a longer transition period. The wrong assessment by the authorities of the UK about the migration potential from Eastern Europe then was an important factor in subsequent events which led to the Brexit.

Prime Minister Yitzhak Rabin was one of those decision makers who used to look with great mistrust at his intelligence agencies. He preferred to be the interpreter and analyst of raw information, sort of saying "you provide me with the information, and I will analyze it myself and reach my own conclusions!"

Much has been said and written about the emergence of the Oslo Accords. Many have also pondered what factors caused Yitzhak Rabin not to inform either the different intelligence agencies or the heads of the defense establishment about his political moves vis-à-vis the talks that were held with PLO officials, which ultimately became known as the Oslo Accords. That Rabin kept them out of the loop is undeniable.

To understand why Rabin made this choice, one must first, in my view, understand the context in which he operated, his outlook, his historical experience, and—perhaps most important—his attitudes toward the different figures who represented key positions at that time.

The historical perspective provided by the lessons of the Egyptian Israeli peace, and by the precedents the Right had created, made Rabin's task easier. In the summer of 1992, there was much innocent enthusiasm among the leaders of the Labor Party. The return to rule was intoxicating. The expectations of Rabin were great. The public and most of the politicians, who were aware of the weight of his responsibility, believed that this time he would be able to meet all the great challenges confronting Israel.

Like many of Israel's leaders, Rabin was a loner. His obsessive suspiciousness toward the people around him rendered him a solitary person. Rabin was well familiar with Israel and with the world at large. He observed events from the perspective of someone who had already undergone many unique and stirring experiences. Hence, he exuded a confidence that he could easily rely on his own judgment, responsibility, and rich experience. He thought deeply within himself and the solutions he offered were a product of that solitude. He did not, however, see himself as someone living inside a bubble, but as someone who always had his feet on the ground. He remained faithful to the reality in which he had developed and functioned, and the solutions he presented were all an outcome of direct learning and a search for truth.

Rabin was not an intellectual and he felt not a little disdain toward those who flaunted themselves as such. Unlike, for instance, Shimon Peres, he never preened before any audience regarding a book he had read, whether it was fine literature, political thought, philosophy, or poetry, and if he read, he did not talk to anyone about it. Nevertheless, as prime minister, Rabin read a good deal of intelligence documents that were sorted for him both by his bureau chief and his military secretary. He would closely peruse every document that was passed on to him. The fact that it had reached his desk meant it was of importance, a message the prime minister needed to know. Generally, he would sign each document that was conveyed to him and record a date beside the signature. Sometimes he added a positive or negative assessment, sometimes directives, and sometimes a request for a discussion or consultation on a matter brought to his attention.

Rabin especially loved raw material and loathed written assessments. His experience with intelligence assessments had been disappointing, and often enough he dismissed them. More than he relied on intelligence researchers, he trusted his own analytical and judgmental ability. He had respect for major figures in intelligence—for example, former head of Aman (Military Intelligence) Aharon Yariv—but went solely by his own assessments and judgment. At the same time, it was important to him to hear the opinions of key officials in the security, military, and defense establishment. This did not necessarily mean he would accept their opinions, but it was a way of testing his own assessments. He put to the test all the senior figures of the security-military establishment who were directly subordinate to him, wanting to hear with his own ears whether their opinion was influenced or derived from a political, sectional, factional, or political-party outlook. Information, he believed, was supposed to be detached from personal influences and to be provided in totally pure form.

As prime minister, he received a constant, voluminous flow of information from open and covert sources, and given his experience and insight, he could immediately tell the wheat from the chaff. Undoubtedly the written and broadcast Israeli media played an important role in formulating his opinion. Rabin would devote special attention to articles and reports covering major events in the world at large. He was not interested in articles by publicists, except for a few elite journalists whom he especially esteemed and appreciated. Clearly, he did not need a newspaper to know what was really happening in the headquarters of Yasser Arafat, with the president of Lebanon, or the president of Syria; nor did he need a newspaper to keep up to date on what was happening in the White House or in the Washington political arena.

In his role as defense minister, Rabin was a different person. Before every meeting or trip aimed at forging security ties, dealing with security exports or imports, he would get briefings orally and in writing from the top foreign relations official of the Defense Ministry, the director-general of the Defense Ministry, and from others engaged in assessment. Before each meeting he always requested that the diplomatic and political ramifications be included in all the briefings, even those that only concerned technological matters.

Despite the importance of his unmediated learning about the international and regional environment, Rabin's database relied first and foremost on the entities working under him. These were orderly and disciplined in their functioning, while competing in the manner of all the governmental entities in the world. They constituted the prime minister's ongoing link with the outside world, and there was no substitute for them.

Rabin relied on himself, on his own judgment and analytical ability. He felt sufficiently sure of himself to forgo the presence of experts, especially those who had not been with him during his long political career. Moreover, as defense minister, Rabin was entitled to all the advisory services within the military establishment along with those of the Mossad and the Shabak (Israel Security Agency), which he was directly in charge of as prime minister.

That was how Rabin operated. It was hard to reach him, hard to win his trust, and hard to persuade him or to dissuade him from things he had made up his mind about in advance.

It is not clear whether, in the summer of 1992, Rabin's assessment was that he would reach a peace treaty with Jordan or shake the hand of Arafat on the White House lawn in Washington. Many have tried to explain the dramatic change that occurred in Rabin on the issue of talks with the PLO, and his famous handshake with Arafat on the White House lawn, by saying he was dragged along by Shimon Peres. Undoubtedly there is a grain of truth in this view. At the same time, it bears emphasizing that from his first day as prime minister, Rabin was convinced of the necessity of first reaching a settlement with the Palestinians.

This is the place to note the special contribution of the head of Aman, Gen. Uri Saguy, who assumed the position in 1991 and took a very clear stance in favor of a peace process with Syria. He claimed that such a process would eventually ensure a settlement with Lebanon and create a ring of peace around Israel. Saguy believed that negotiations with the Palestinians were pointless. In each briefing of the prime minister, he made sure to emphasize the tumult among the Palestinians, the wars between personalities, the confrontation between the "inside" and the leadership in Tunis, Arafat's lack of credibility, and so on—in contrast to the order and discipline prevailing in Syria. In Syria there were "those you could trust." President Assad had proved that he could uphold the 1974 Separation of Forces agreement between Israel and Syria. In brief, in Damascus there was an address, whereas in Tunis there was no one to be trusted.

The wave of terror perpetrated by Hamas at the end of 1992, along with the expulsion of 415 of its operatives to southern Lebanon, led to the end of the peace talks in Washington. Additional terror attacks carried out at the beginning of 1993 led Rabin to declare a long-term curfew in the territories, a step that brought about a freeze in the talks with the Jordanian-Palestinian delegation. In July 1993, Operation Accountability effectively put an end to the Madrid formula he had inherited from the previous Shamir administration and which he disliked profoundly.

That may have been what Rabin was striving for from the start. He never felt comfortable with the Likud legacy for the format of the negotiations that was determined in Madrid. He did not like it but could not unilaterally abrogate it. He did not believe it would ever be possible to make progress when all the Arab participants appeared together for talks. He claimed that in such a situation, no

Arab side would dare to make progress for fear of being accused of treason. Each Arab side would monitor the other, and this was an ideal prescription for a lack of progress and for merely marking time.

Rabin thought differently and adduced, for example, his personal experience as a member of the Israeli delegation to the 1949 armistice talks that were conducted on the island of Rhodes after the War of Independence. More than once he pointed to the fact that the logical prescription for progress consisted of step-by-step negotiations with all the Arab states.

Apparently, the chiefs of intelligence and of the defense establishment did not pay heed to an interview Rabin gave to the IDF weekly Bamahane on September 29, 1992, in which he said:

> When I was defense minister in the National Unity Government a different format for the peace process was used. Whoever looks into the history of the relations between the Arab states and Israel since the War of Independence will find that in none of the international summits did we achieve peace. Even if there was a limited agreement, like the armistice. Sitting opposite you is a person who was a member of the delegation to the first negotiations between the state of Israel, which was not yet a year old, and Egypt. We sat in February and March 1949 and reached an agreement. We finished with the Egyptian delegation, we sat with the Jordanians. In the negotiations with the Syrians, we no longer needed to go to Rhodes. One meeting was held on the Israeli side, at Rosh Pina, and a second meeting—at the Syrian customs house, which at that time was in the Syrian part of the Golan Heights; the talks with Lebanon were held in Nakura. That is, there too we did not have to remove ourselves from the region.
>
> After that there was a 25-year drought; the Sinai Campaign and the Six-Day War did not end in agreements. Only after the Yom Kippur War was there a return to agreements. We signed a Separation of Forces agreement with Egypt, and after that we began negotiations with Syria. Then came the Sinai Interim Agreement, then Camp David, and then the peace treaty was signed.... Since 1979 no agreement has been signed except a stillborn one on May 17, 1983, the peace treaty with Lebanon that was already null and void before the ink of its signers had dried.
>
> The Madrid Summit is an international-conference format with all the shortcomings entailed. The three Arab delegations—Jordanian-Palestinian, Syrian, and Lebanese—come together at the same time and the same place. Because this situation creates a linkage that spells danger, the format is not ideal. It was pressure on the Shamir government that produced the format, and I inherited it. There is no doubt that it is not ideal. Therefore, we insist that each negotiation with the Arabs stand by itself.

Rabin asserted that only clandestine contacts, far from the media spotlight, could lead to the long-desired agreements. He would reiterate countless times that the peace with Egypt could not have been reached without the prior meetings between Moshe Dayan and Anwar sedate close adviser, Hassan Tuhami in Morocco. The same held true for Oslo and for the peace treaty with Jordan. Moreover, Rabin believed that without intensive American involvement no peace negotiations would reach their goal.

Both in private conversations and in open declarations, the prime minister would point to the peace agreements with Egypt as an example of the advantages of separate or step-by-step negotiations, and to the complexity that characterizes all negotiations, requiring secrecy. He would emphasize especially that President Carter had to come personally to Cairo, to Jerusalem, and back to finalize all the understandings that were reached between Israel and Egypt before the peace treaty was signed in Washington on March 26, 1979. Without Carter's personal involvement, it is doubtful whether the gaps between the sides could have been bridged, even though Israel had agreed from the start to return the entire Sinai Peninsula to Egypt.

On December 29, 1992, Yitzhak Rabin gave a speech in which he said, among other things:

> If the U.S. president had not invited Mr. Begin and President Sadat and isolated them at Camp David for 13 days without media, they would not have come out with the same Camp David Accords, and this was despite all the crises that occurred. Six months after the Camp David Accords, when he had to come for the signing of the peace treaty, the U.S. President flew by himself on Air Force 1 to Cairo and from there to Jerusalem and back. The agreement, of course, was not signed either in Cairo or Jerusalem but on the White House lawn in Washington.

Nevertheless, Rabin could not abandon the Madrid formula without an alternative one, and certainly could not have allowed himself to do so from a public and political standpoint. He was especially concerned about the issue of Palestinian representation. Any attempt to jettison the Madrid formula could have raised the issue of PLO participation in direct negotiations, a problem that was circumvented, with US agreement, before the Madrid Summit in October 1991. After ten rounds of talks, it was clear to everyone that it was impossible to continue in this manner, and certainly not at that level.

It was President Assad himself who offered Rabin a way out of the cul-de-sac. It was he who proposed to the US emissary Dennis Ross those negotiations be held "in the Henry Kissinger style"; he thereby signaled to Rabin that he was prepared to negotiate at the most senior level and with full US mediation. Assad was aware of Rabin's intentions and knew that only in that way was there perhaps a possibility to reach some compromise. Rabin had a special attitude toward Hafez al-Assad. He esteemed him as a leader who kept his word; like the military and intelligence chiefs, he assessed that in Syria "there is an address." It was not for nothing that he often reiterated that the separation agreement with Syria had been maintained without any Syrian violation, making the border with Syria the quietest of Israel's borders for more than 18 years.

Beyond all those factors, however, it was the contacts with the Palestinians in Oslo that really made it possible to abandon the Madrid formula. This was even though, at first, those contacts were not taken seriously by Rabin himself.

On Thursday, July 29, 1993, US Secretary of State Warren Christopher called Rabin and told him he had decided to visit the region. With the secretary of state's visit approaching, Rabin decided to hold an urgent meeting at his bureau in Tel Aviv. The prime minister asked to hear the intelligence assessments first. Saguy gave an overview and explained that there was turmoil in the Palestinian camp and no chance to reach a settlement with them. The head of Aman pointed clearly in the direction of Syria. He added with a half-smile: "But you know well what must be decided. I cannot be in your place and know what you will do, but you are aware of Syria's positions."

The prime minister expressed distrust toward the Palestinian issue and said it would not hurt to make that attitude clear to the Americans. Chief of Staff Ehud Barak, too, came out in favor of dealing with Syria first, saying it would afford a rare opportunity to change the situation in the north and in Lebanon. The prime minister asserted that the Palestinians would start to "move" only once they realized that the Syrians were progressing on their own front, and said he agreed with Aman's assessment of the Palestinian issue. Rabin thereby expressed his frustration with the Palestinian track.

The beginning of August 1993 was also a landmark for the Palestinian camp. Already during Christopher's visit, details emerged about sharp disagreements between PLO-Tunis and the Palestinian delegation to the talks. As Rabin described the situation to Christopher: "There's a mess there that will be hard to get out of."

When I entered his office on the morning of August 16, 1993, the prime minister pulled out a green envelope and drew from it two printed white pages, one in Arabic and one in Hebrew. "I am told that this arrived from Abu Mazen [Mahmoud Abbas]. Translate it and bring it to me."

The document in Arabic that the prime minister gave me included these five paragraphs: The two stages: We agree to two stages—a transition stage and a final stage. The final-status issues will be defined within the framework of the Declaration of Principles so that the two stages can be linked: Jerusalem, refugees, settlements, borders, and so on, in accordance with the agreement. Jerusalem: We agree that the status of Jerusalem will be discussed as part of the final settlement. Now: participation in elections, to elect and be elected. Linkage between the Jerusalem institutions and the temporary authority, and supervision of them.

The settlements: We agree that their status will be discussed in the final stage. For now, they will remain under the Israeli army's responsibility as the transition stage. A special agreement will address this issue at the transition stage because of its difficulty and complexity.

Security: External security is under Israel's responsibility during the transition period and internal security is under the Palestinian Authority's responsibility. A liaison and coordination committee will be formed to solve the common problems and the disagreements.

We think any agreement on a Declaration of Principles must be accompanied by the taking of a big, significant step between the Palestinian organization and the Israeli government, before the agreement and the signing of the declaration.

When I finished the translation, I brought the document to Rabin. He smiled, folded it into the green envelope, put it in a drawer of his desk, and again told me not to speak of it to anyone. At that moment I told myself that Rabin had gotten what he wanted and from now on the way to an agreement with the PLO would be open. Until then there may have been talks and understandings with the PLO, but only on August 16, 1993, a month before the signing on the White House lawn, did the constellation emerge that enabled Rabin to cross the political Rubicon, leave aside the Syrian track option and favor an agreement with the PLO. Undoubtedly, up to that point, Rabin had known about the activity of Peres and his team, but only on August 16 did he decide to make his contribution to reaching understandings with the PLO. From a certain standpoint, that day of August 16 symbolizes Rabin's transition from an uncommitted observer (perhaps even a nonbelieving observer) to a key and decisive actor in the negotiations with the PLO.

Still, at the government meeting of August 22, 1993, Rabin declared that it would be a mistake to stop the negotiations with Syria. Rabin stressed that it was impossible to demand that Syria promise to stop the terror in Lebanon as a condition for continuing the talks; the Likud government before him had not done so. Therefore, the Lebanese problem remained as it was: "painful, threatening, and unsolvable."

On August 25, 1993, Prime Minister Yitzhak Rabin revealed his moves to the representatives of the Presidents' Conference, Malcolm Hoenlein and Lester Polack, particularly regarding the Palestinians and the PLO. The two first wanted to know whether the president of Syria, Hafez al-Assad, could be trusted at all, considering that he was preventing the Lebanese army from deploying in the south, encouraging Hizbullah to attack Israel, continuing to transfer weapons and ammunition from Iran through Syria to Hizbullah, and acquiring ground-to-ground missiles from North Korea.

Rabin went on to explain that as far as he was concerned, and as he had made clear to Christopher, he would not be able to make a decision that would include both tracks together, the Syrian and the Palestinian. He would not be able to manage that politically; instead, he would choose one of the tracks as soon as there was a chance of a breakthrough. When Hoenlein and Polack asked for his position on the "Gaza and Jericho first" option, Rabin explained that as he saw it, there were two problems: Jerusalem, which he was in no way prepared to concede, and the question of the location of the Palestinian Authority that would be established following the agreement on self-rule.

He would not agree to the Palestinian Authority being in Gaza or in Jericho. He then added: "Possibly in Ramallah." But not Hebron, because it was a city with a significant Jewish presence. Rabin added that the PLO's situation was difficult; it had internecine problems, leadership problems, and financial problems. "Maybe, now that they're down, they'll agree to give up some of their demands," the prime minister remarked. Later in the candid conversation, Rabin admitted that there was a package deal on the agenda in which he would agree to include Jericho in the Gaza-first agreement if the Palestinians were to concede on the Jerusalem issue and the location of the Palestinian government.

Those were not all of Yitzhak Rabin's disclosures. He noted that the PLO was now prepared to pay a very heavy price to be recognized by Israel. Hoenlein, taken aback, asked in a constricted voice whether the prime minister was really considering such an option. Rabin replied: "There are many conditions, such as stopping the terror, changing the Palestinian Covenant that includes a call to destroy the Jewish state, and so on."

On August 26, 1993, Rabin held a briefing for the heads of the Israeli delegations to the talk. It was just before the resumption of the eleventh round in Washington. In typical fashion, Rabin opened the discussion and asked Gen. Saguy for the intelligence assessment. The Aman chief began by asserting that "Assad has crossed the

Rubicon," though he immediately added: "That does not necessarily mean there will be progress at this point." Saguy said Assad knew he could no longer suffice with the same stock phrases; he would have to meet the full price that Israel demanded, which included, among other things, diplomatic relations. "Assad decided so himself."

The Aman chief added that unlike in the past, Assad was prepared for progress in the Lebanese arena, even before a Syrian Israeli agreement. He stressed that Syrian figures had referred to the need "to accept what is hard to accept," thus hinting at Syria's readiness for normalization with Israel. Saguy remarked that despite his lack of knowledge of Arabic, he was following the Syrian media and it showed a moderating trend on the issue of peace with Israel. "Experts in Arabic" had told him their impression was that Assad was seeking a comprehensive settlement "and it does not matter to him whether it will involve the Palestinians or another arena. Assad will not wait for the Palestinians indefinitely." With this the Aman chief ended his statement on the Syrian issue.

Saguy went on to describe the Palestinian arena and likened the latest developments there to an "earthquake." The struggle between the "inside" and the "outside" was having its effect. Some understood that Arafat had already decided and that his political future was at stake. True, there was no substitute for the PLO "and Arafat has lost status," but still the Aman chief believed Arafat would prevail. "Those who are fleeing the ship understand what he has done." At the same time, Saguy repeated his familiar thesis that even though Arafat would win out, in the struggle between the "inside" and the outside" the former was gaining strength.

The prime minister did not bat an eye. He did not say a word about the deal that was taking shape with the PLO, and he gave an exaggerated assessment of the results of the coming round, as if that track were the only one that existed. Rabin declared: "If this round ends with nothing, there will be a very serious problem for Arafat!"

At a certain point, Rabin asked the participants what was more important: to first reach a settlement with Syria or with the Palestinians? The Aman chief replied immediately that it was preferable that Syria come first but said this would entail a price for Israel. Rabin answered that he was seeking his opinion and not that he should respond in his stead, "not at this stage at least."

Rabin added that the strategic advantage in reaching peace with Syria was clear, and the price was clear as well. However, while what would be achieved with Syria would be irreversible, what would be achieved with the Palestinians would be reversible; it would involve territories adjacent to Israel in which Israel's ability to intervene was beyond question.

Here Rabin clearly indicated that he preferred the Palestinian option. The discussion concluded without decisions being made and without directives to the heads of the delegations. Rabin rose from his chair with the statement: "I do not envy whoever will give back the Golan!"

It was clear to the participants that Rabin was not disclosing any secrets to this forum regarding the "Gaza and Jericho first" talks. And this was even though the Israeli press had been reporting extensively for days that Foreign Minister Shimon Peres was preparing a "package" that was now taking shape. Nor did any of the participants know about Peres' upcoming trip to the USA with an almost-finalized agreement in his pocket.

In a meeting that the prime minister held with Steve Grossman, then the president of AIPAC. Rabin made use of the occasion to update him on his diplomatic moves. For the first time, Rabin talked about the four months of direct talks between Israel and the PLO.

Rabin went on to say that, during the last four months of contacts with PLO representatives, the PLO had suffered from a lack of budget and was in a state of bankruptcy, and for that reason it was losing ground. Hamas was receiving funds from Saudi Arabia and from the Muslim community in the USA for the charity organizations it sponsored. Whatever would be decided about the Golan was irreversible. However, what was done in the territories would be reversible, and there was no problem regarding Gaza.

The talks had reached the stage of a Declaration of Principles. Jerusalem, the settlements, and the military bases would not be given to the Palestinian Authority. As for the Authority's powers, they would be determined together with Israel. There was a readiness to withdraw from Gaza, while the lines of confrontation (the Jordan River, the bridges across it) would be in Israel's hands. The IDF would protect the settlements and the Israelis in Gaza. Rabin stated: "Wherever there are Israelis, we will be there."

Rabin referred to three levels of the negotiations. The first concerned the implementation of "Gaza and Jericho first." (He did not mention the other two levels at all.) He said Israel would have to take painful measures. The USA would have to help Israel counter the Arab boycott, help it develop relations with moderate Arab states, and give it financial aid so as to strengthen it.

On September 9, 1993, Prime Minister Yitzhak Rabin signed the document recognizing the PLO as the representative of the Palestinian people, and it was finally agreed that the signing ceremony for the Declaration of Principles would be held on the White House lawn. On September 13, 1993, the agreement was signed.

If so, how could it be that Israeli intelligence did not know or was not informed about the Oslo measures? In my assessment, it was for several reasons:

First, the talks at the beginning of 1993 between PLO officials and an Israeli group led by Foreign Minister Shimon Peres were held, in more ways than one, far from the spotlight. They were held in Oslo (sponsored and hosted by the Norwegian government) on several occasions, at the end of which the document known as the Declaration of Principles was drafted. As it turned out later, Rabin was aware of the Oslo talks but did not investigate the details of the agreements being reached and hence was not knowledgeable about them. Presumably, Rabin saw these talks as less important than they were. The Israeli political scene was rife with political figures and academics who, to Rabin's chagrin, held talks with PLO officials. Nothing resulted from all these talks except gossip and intelligence information about what was happening in Arafat's court.

In general, the whole issue of the talks between Israel and the PLO was very irksome to the prime minister. Not only did he know about almost all the channels of dialogue between official or other Israeli actors and PLO representatives, but the prime minister knew details of all sorts of deals that were hatched, of initiatives apparently taken by Israeli actors who said that if the Palestinian delegation were to adopt a certain stance, they would ensure that Rabin would agree to it, and the like. Rabin would scorn these efforts and say that the Israelis involved were wasting their time.

Rabin kept regarding the group conducting the talks in Oslo the same way until he realized that the progress being achieved there was real. From Rabin's standpoint the Oslo talks met two of the three conditions he had posited for the success of any negotiating process with an Arab actor: nothing had been leaked about them and the meetings were with a separate Palestinian delegation, though instead of US sponsorship there was Norwegian involvement. It was a replication of the Dayan-Tuhami meeting in Morocco that had preceded the Camp David Accords.

If so, how to explain the intelligence failure? How did intelligence not detect the talks being held in Oslo? How did the matter not leak to the media? There is no explanation for this. The Israeli Intelligence Community was surprised in 1977 about Sadat's intention to reach peace with Israel, and here again, it was surprised in 1993 on the Palestinian issue.

Possibly the most reasonable explanation lies in the simple fact that the eyes of intelligence, like the eyes of the security and political leaders, were focused on the Syrian track. The Palestinian-Jordanian track did not stir anyone's hopes. The feeling was that it was impossible to get anywhere the Palestinians. The intelligence-collection branches were focused mainly on the Syrian track.

Rabin's zigzags between the two tracks—Palestinian and Syrian—did not make things any clearer for intelligence. Moreover, there is no doubt that Rabin remained committed to his approach and enjoyed obfuscating the issue of the talks with the Palestinians. His gamble was simple: if the talks failed, it would not be possible to blame him, and if they led to a positive outcome, he would be part of the success. Hence it was preferable to keep the whole issue of the negotiations with the PLO in secrecy, and so long as intelligence did not know about it, there was no reason to update it and endanger the existence of the talks through a chance and unintended (or perhaps intended) leak by any of the customers of the intelligence material.

Rabin's personality structure, obsessive innovativeness, the compartmentalization policy he practiced, the dim view of intelligence assessments that he had developed, his tense relations with the Aman chief, and the fact that no one apart from those negotiating in Oslo knew about the talks, encouraged Rabin to continue his policy of concealment. There was, however, one major difference: when Rabin realized that the talks had led to agreement on the Declaration of Principles, he hastened to add the legal adviser Joel Singer to those who were in on the secret and instructed him personally to get involved in the talks. The Declaration of Principles, however, had already been agreed upon. Hence Rabin insisted on adding what he called "Agreed Minutes," or clarifications, to the document. More than once Rabin said that without the "Agreed Minutes" the Declaration of Principles would have been a "national disaster"!

In retrospect, it appears possible that Rabin would have reconsidered his steps if he had met (as he told me and said several times in different forums) with Arafat before the signing of the Declaration of Principles. Rabin was not well-versed in the negotiations that were held in Oslo. This became clear to me several times when he asked me to investigate what had been said in Oslo and whether promises had been made that Arafat had flaunted so as to cut corners and reach an agreement. Both the advantage and the disadvantage of the Declaration of Principles lay in its ambiguity. Rabin did not know Arafat's real positions, only what had told him by those engaged in the Oslo talks. Several times I heard him say that without the "Agreed Minutes," which were an outcome of his direct intervention, the Declaration of Principles that was reached with the PLO would have been a national disaster.

On the critical point of the negotiations on the implementation of the Declaration of Principles, Rabin said that he did not have the information he needed on the PLO chairman's positions to help him decide on the issues at stake. The question that is asked, therefore, is whether intelligence could have supplied him with the information he lacked before the Oslo Accords were signed. The answer is: maybe yes, maybe no. Eventually Rabin received the information he had lacked through direct contacts with Arafat after the signing of the Oslo Accords, and particularly in the long meeting between the Israeli prime minister and the PLO chairman in Cairo in December 1993, during which Rabin was directly informed of Arafat's positions. The problem was that Rabin did not give intelligence an opportunity to disprove Arafat's claims.

As in the talks that preceded the peace with Egypt, in the Palestinian case, too, intelligence remained outside the initial meetings that later developed into the Oslo Accords. True, Rabin chose not to inform intelligence of his considerations, but intelligence bears some of the blame as well. The basic shortcoming of intelligence remained the same as before the 1973 Yom Kippur War: difficulty in assessing leaders' intentions. The inclination to prefer the Syrian track and to disdain the Palestinian camp, describing it as not knowing what it wanted, is an indication of a conception that had rigidified and had all the elements of an assessment failure.

This article was first published in the JPSR (Jewish Political Studies review). It was re-edited for the purpose of this book.

Tausch on Terrorism

Hussein Solomon

Abstract It is indeed difficult to capture the essence of a lifetime of scholarship of a scholar as prolific as Arno Tausch. This is made even more difficult given his varied interests. Rather than focus on his numerous works examining terrorism in its myriad manifestations, this chapter seeks to examine Tausch's stance on terrorism in three areas: what drives recruitment into terrorist organizations, the relationship between Islamism, terrorism and anti-Semitism and what is to be done to minimize the scourge of terrorism? Tausch's own works are contextualized within existing scholarship on terrorism. It will be argued that his greatest contribution to the study of terrorism lay in his wide-angled approach, incorporating both qualitative and quantitative aspects.

1 Introduction

According to the Global Terrorism Index, more than 170,000 terrorist incidents have occurred between 1970 and 2019. More to the point, much of the violence have moved away from the anarchist, nationalist and leftist waves of violence and has taken the form of militant Islamist violence. More specifically, 160,000 lives were lost as a direct result of jihadist violence between 1979 and 2019 (Institute for Economic and Peace, 2020, 2). This, in turn, has served as a catalyst for much research on the subject.

To put it differently since 9/11 one new book is produced on terrorism every six hours. Despite the tomes produced, despite the conference papers on the subject, old questions remain unanswered. Marc Sageman (2014, 565, 569) opines, "Despite a decade of government funding and thousands of newcomers to the field of terrorist research, we are no closer to answering the simple question of 'What leads a person to political violence?' . . .[A]fter all this funding and this flurry of publications, with

H. Solomon (✉)
Centre for Gender and Africa Studies, University of the Free State, Bloemfontein, South Africa
e-mail: solomonh@ufs.ac.za

G. Segell (ed.), *Development, Globalization, Global Values, and Security*,
https://doi.org/10.1007/978-3-031-24513-8_15

each new terrorist incident we realize that we are no closer to answering our original question about what leads people to political violence".

What accounts for this failure in understanding terrorism? Part of the problem is that much scholarship focuses on it at the expense of attempting to study the phenomenon more holistically. Whilst some studies examine Islamic State, others focus on the likes of West Africa's Boko Haram. This compartmentalized approach becomes difficult to sustain when one sees growing cooperation between different terror groups resulting in their merger. Islamic State West Africa Province reflects this trend where members of Boko Haram act as the local franchise of Islamic State. Other scholars examining terrorism focus on a particular region. However, in an era of globalization where insecurity anywhere threatens security everywhere such a regional focus is problematic where groups like Lebanon's Hezbollah has moved out of national or regional territorial confines. Indeed, their operations span the Americas, Africa and deep into Asia (Solomon, 2017).

Other academics focus on thematic themes—issues of radicalization or counter-insurgency or a focus on counter-terror legislation. Terrorism, of course, covers all these issues but the thematic approach to the study of terrorism emanates from the disciplinary homes that scholars emanate from. A focus on counter-terrorism legislation is a natural basis for a legal scholar when examining the phenomenon, a sociologist may well focus on issues of socialization and radicalization whilst a strategic studies scholar might well focus on counterinsurgency.

Whilst providing useful insights into disparate aspects of terrorism, what is needed are more holistic approaches that can provide insight into the origins and formation of terrorist groups as well as the strategies and tactics they adopt and one which also provides analysis on the effectiveness of counter-terrorism responses from legislative frameworks to intelligence and military responses. The reason is obvious: a narrow focus on issues of recruitment without understandings of the underlying structural motivations promoting popular resentment and radicalization is bound to result in failed counter-terrorism policies.

In the search for a more holistic approach to understand terrorism, what is needed is a Renaissance-type scholar who can transcend disciplinary boundaries and one who whilst zoning in on the particular of the phenomenon is aware of how broader dynamics in state and society impact on the evolution and trajectories of militant groups. Arno Tausch is such a scholar. Whilst securing his Ph.D. in Political Science, he is equally adept in sociology, theology or economics. His research has covered areas as diverse as child poverty, dependency theory, globalization, arms exports to developing countries, anti-Semitism, Kondratieff's long economic cycles, migration, Catholicism, Islamism and terrorism.

Moreover, his work transcends geographical boundaries—whether his native Austria, Brazil, Russia or the Middle East—the world is Tausch's laboratory (Tausch & Prager, 1993; Tausch, 2001, 2009, 2010, 467–488; Tausch et al., 2015, 543–554). Tausch also straddles both qualitative and quantitative approaches to the study of terrorism. This merges the critical depth of qualitative approaches with the more expansive scope that quantitative approaches offer. This, in turn, allows him an opportunity to engage in far more incisive comparative insights than his peers.

2 What Drives People into Terrorism?

This is the question Marc Sageman posed (above) on the limited utility of much contemporary scholars on terrorism. In his various articles, Tausch has also reflected on the same issue. In his Social Attitudes Fueling Islamist Terrorism, Tausch (2021, 187–218) deployed a multivariate methodology and data from the original IBM-SPSS XXIV evaluation of the Sciences Po Paris database Islamist Terror Attacks in the World, 1979–2019. This database provided Tausch unique insights into the perpetrators, affected institutions and 160,000 victims of Islamist terrorism over this 40-year period.

This data was then correlated with PEW's dataset The World's Muslims: Religion, Politics and Society, the Arab Barometer, Arab Opinion Index, and World Values Survey. The latter is especially important on account that it covers 72 countries of the world, representing 80 percent of the globe's population. These datasets were then contextualized with primary documents such as the United Nations Development Programme's (UNDP) Human Development Reports and extensive secondary literature. The meticulous and rigorous nature of the methodology demonstrates Arno Tausch at his best as a leading scholar on terrorism.

His insights go beyond traditional responses to terrorism. In this chapter, the close connection between restrictive gender norms, Islamist sharia ideology, and terrorism is explicitly made. For instance, utilizing datasets from Pew and the Gender Social Norms Index, he demonstrates, "...that the two main drivers of Muslim support for suicide bombing are the advocacy of honour killings and the death penalty for Muslim apostates" (Ibid. 188).

His research finds resonance with other scholars. Annela Salman's, for example, persuasively argues that, "...females actual advancement and equality in education, employment and political representation are more effective in reducing terrorism" (Quoted in Freilich et al., 2015, 278). The Gender Social Norms Index which tracks gender bias and obstacles to gender emancipation focuses on four areas: politics, education, economic/work and physical integrity (United Nations Development Program, 2022).

In Muslim majority countries ranging from Afghanistan and Pakistan to Iran, the Palestinian territories, Nigeria and Yemen, Turkey and Tunisia, gender bias is most prominent. This prominence of gender bias also has its corollary in that in many of these same countries, majorities of Muslim public opinion support the implementation of sharia law. As Tausch highlights, 99 percent of Muslims support sharia law in Afghanistan, and 84 percent of Pakistani Muslims support full implementation of sharia law (Tausch, 2021, 190).

It should come as no surprise that Afghanistan and Pakistan are also plagued with terrorism. Tausch's comprehensive approach is in keeping with Jerrold Post's critique that explanations of terrorism at the individual level are insufficient in trying to understand why people become terrorists. Rather, Post (2007, 8) provocatively asserts, "... terrorists are not depressed, severely emotionally disturbed or crazed fanatics. It is not individual psychopathology, but group, organizational and social psychology, with a particular emphasis on collective identity that provides the most powerful lens through which to understand terrorist psychology and behavior".

In keeping with Post's refrain, Tausch's approach to terrorism is a wide-angled one, situating the phenomenon within a wider context. In this more holistic approach, Tausch is following the example set by Mark Tessler's path-breaking work on terrorism which views its origins also in attitudes towards democracy and attitudes towards western culture and society, including gender norms (Tessler, 2015). Explaining his adoption of this wide-angle lens, Tausch (2021, 191) eloquently argues, "...it is wrong to define radical Islamism and also ongoing radicalization of such mass organizations as the Muslim Brotherhood or the Turkish Milli Gorus, which both start, like the most radicalized factions of Islamist terrorism, from the intense hatred of "Jews and Free Masons" and Western Civilization as such, and which for many appear as "moderate Islamists" and worthy partners of dialogue, while in reality they provide the fertile ground from which armed terrorist groups only can develop".

The utility of Tausch's more holistic approach to terrorism provides us with the understanding to connect so-called moderate Islamism with more radical Islamists. The case of Jamaat al-Tabligh wal-Da'wa (Society for Propagation and Preach) illustrates the point well. They play an important role in the radicalization and recruitment of Muslims for the Islamist cause. The activities of the organization, its membership or finances are never published, and officials of the organization assiduously work to remain outside both media and government attention. This secrecy is further enhanced by the leadership's dynastic character. All Tableegh Jamaat leaders since Maulana Muhammed Elyas Kandhalawi (1885–1944), its founder, have been related to him by either blood or marriage (Alexiev, 2005, 3–4). Ideologically, Tableegh Jamaat hardly differs from other radical Islamists with its ultimate objective being nothing short of a "planned conquest of ethe world in the spirit of jihad" (Ibid., 4).

Indeed, over the years, the 'quietest' and non-political image of Tableegh Jamaat has increasingly been questioned as events contradicted the image. In 1995, there was an attempted coup in Pakistan that involved high-ranking military officers and civilians—all of whom were members of Tableegh Jamaat. Kazakh authorities, meanwhile, deported several the organization's members for "advancing extremist propaganda and recruitment" (Ibid., 6).

In Morocco, meanwhile, another Tableegh Jamaat member—Yusuf Fikri—who was also the leader of the Moroccan terrorist group, At Takfir al-Hijrah, was sentenced to death for his role in the May 2003 Casablanca bombings which killed more than 40 people. In the Philippines, the government accused Tableegh Jamaat of serving both as a conduit of Saudi money to the Islamist terrorists in the south and for Pakistani jihad volunteers to join in the insurrection in the south (Solomon, 2013, 24).

In France, 80 percent of young Muslim extremists emanate from Tableegh Jamaat's ranks, prompting one French intelligence official to refer to them as the "antechamber of fundamentalism" (Alexiev, 2005, 5). Imam Ben Halim Abderraouf, based in Paris and a senior figure within Tableegh Jamaat, played a key role in recruiting young Dutch Muslims. At least four of them then travelled on fake Algerian passports to Damascus via Paris and then proceeded to the Syrian-Iraqi border to receive military training from Iraqi insurgents (Block, 2005).

This fear about Tableegh Jamaat was not only expressed in Europe but also in Argentina when twenty-six of their members entered Buenos Aries. These members came from a diverse number of countries, including Pakistan, Malaysia, Qatar, Egypt and South Africa. Security services from Spain, Italy and Argentina expressed their fear that their aim was to recruit Argentine Muslims to be trained outside in terrorist activities (Hassan, 2005).

Spanish anti-terror experts were also investigating some of the group's members for involvement in the Madrid train attacks on 11 March 2004 which killed 193 people and therefore provided movements on this group of twenty-six to the Argentine authorities. In December 2021, Saudi Arabia, long a supporter of the group, condemned the group, saying they opened the gates to terrorism (Xulu, 2021).

India's ruling Janata Bharata Party (BJP) endorsed the Saudi position (Hindustan Times, 2021). The attack on a Texas synagogue in January 2022 by Malik Faisal Akram once again highlighted the Tableegh Jamaat connection. Akram had travelled 11 times to Jhelum in Pakistan between 2007 and 2020 with other members of Tableegh Jamaat of which he was also a member (Marino 2022). As Gelb has pointed out Tableegh Jamaat has often been serving as the front door for Al Qaeda (Hassan, 2005). According to Al Alexiev (2005, 5), recruitment for young jihadists is identical:

> After joining Tableegh Jamaat groups at a local mosque or Islamic centre and doing a few local dawa (proselytism) missions, Tablighi officials invite star recruits to the Tablighi centre in Raiwind, Pakistan, for four months of additional missionary training. Representatives of terrorist organizations approach the students at the Raiwind center and invite them to undertake military training. Most agree to do so.

Tausch's holistic approach to terrorism therefore provides one with deep insights connecting moderate Islamism with radical Islamism, one which policy makers can therefore follow in order to mitigate, if not eradicate, the phenomenon of Islamist terrorism. The pressing need for more holistic solutions is also necessitated by the widespread sympathies displayed by the world's Muslim population towards terror groups and their cause. In his incisive article Estimates on the Global Threat of Islamic State Terrorism in the Face of the 2015 Paris and Copenhagen Attacks, Tausch (2015) draws on the various Pew Surveys of the Muslim world which pointed out that 17.38 percent of Muslims globally hold a favourable public opinion for the likes of Al Qaeda, Hamas, Hezbollah and the Taliban as well as for the act of suicide bombing. Tausch emphasizes that if one considers that the global Muslim population is 1.9 billion, it means that 330 million Muslims hold some quite extreme views. This suggests that in addition to holistic solutions, global responses are needed.

3 Islamism, Terrorism and Anti-Semitism

Anti-Semitism, Tausch opines, is a vital component of Islamist ideology driving extremist Islamist movements (Tausch, 2014, p. 36). Reinforcing the work of Bassam Tibi (2002, 2012), Tausch notes that latter day Islamists believe that a Jewish conspiracy against Islam did not begin with the founding of the state of Israel, but rather began at the very origins of Islam in 622 (Tausch, 2021, 191). This virulent anti-Semitism is evident in a perusal of Osama bin Laden's tract Jihad Against Jews and Crusaders (Laqueur, 2004, 412). Bin Laden's successor as Al Qaeda leader, Ayman al-Zawahiri, penned his own diatribe Knights under the Prophet's Banner (Ibid., 427–428). In this tract, the enemies of Islam, included not only Jews but the United Nations, multinational corporations, international news agencies, international relief agencies as well as international communication and data exchange systems.

Underlying their worldview is a global Jewish conspiracy wherein, Jews are seen to be controlling world affairs—political social and economic (Solomon, 2013, 14). Perhaps, the best exemplar of this perspective is the Hamas Covenant. Article 22 tells the reader that Jewish capital controls much of the world; that the Jews were behind the French Revolution of 1789, the Russian Revolution of 1917, as well as the two World Wars. Moreover, we are told that the Jews had set up freemasonry, Rotary Clubs and Lion Clubs as destructive spying agencies (Laqueur, 2004, 427–428).

It is therefore quite fitting for Tausch to examine the phenomenon of anti-Semitism. In the Middle East North Africa (MENA) region, with its majority Muslim demographics, Tausch reminds us that 74 percent of the population display some anti-Jewish bias from not wanting to have a Jewish neighbour to a deep sense that Jews constitute a source of existential insecurity (Tausch, 2020, 126). This would suggest that the various extremist organizations operating in this restive region, may well represent public opinion better than the regimes in the region. What this also suggests is that the various Arab regimes who signed the Abraham Accords with Israel may well be out of step with public opinion (Segell, 2021, 243–246). Given the fact that these are autocratic regimes, it raises a thorny question for policy makers: support authoritarian regimes who are moderates and suppress the democratic impulse or encourage democratization and risk Islamist regimes coming to power via the ballot box as had happened in Egypt following the ouster of Mubarak?

4 What Is to Be Done?

Tausch's numerous publications suggest that Islamist terrorism will not merely evaporate as a result of appropriate legislative frameworks or effective counterinsurgency (Solomon & Tausch, 2020, 2021). Rather, what is needed is a much deeper transformation of Muslim society starting with their recognition that Islamism is fundamentally contrary to the teachings of the Qur'an. He stridently argues,

> We should make it clear here that the Islamist terror ideology and practice are against everything which Muslim humanism stands for, namely a successful and just economic system, good governance, respect for human rights, freedom of choice and protection under a specified code of law, and a code by which to manage external relations (with non-Muslims). . . .The Qur'an—contrary to the reading by the Islamists—considers humans to be the crowning achievement of Allah's Creation for whose personal and collective development everything else has been created. Humans are endowed with intelligence to know their Creator, to recognize and appreciate the universe and everything in it and to understand the reasons for their own existence as contingent on the Will of their Creator. Given the dignity that Allah has bestowed on humans and the sanctity of Allah's creation—humans—all humans must respect the dignity and human rights of others. To harm a human is to harm all humanity (Tausch, 2021, 203, 207).

Tausch's approach for the transformation of Muslim society is informed by the fact that terrorism cannot be rooted out without addressing the structural conditions and social milieu which is producing violent extremism, from authoritarian regimes to closed economies and societies characterized by gender discrimination (Solomon & Tausch, 2021). This led him to spend considerable time examining the prospects of a global Liberal Islam (Tausch, 2021, 208–211).

His reflections mirror that which is taking place in much of the Muslim world led by Scheherazade Rehman and Hossein Askari (Rehman & Askari, 2010, 1–37) who have sought to create an authentic Islam which was compatible with liberal democracy, Bayat (2005) who has coined the phrase post-Islamism to refers to an inclusive religiosity which reinforces liberal democratic norms as well as Komecoglu (2020) who speaks of a civil Islam which focuses on an individual Muslim's spiritual development and the general welfare of society in juxtaposition to Islamists' obsession with political power.

5 Conclusion

Terrorism remains an ongoing concern for all societies. The case was made that it is imperative to approach it holistically, examining terrorism within the context of economic stagnation, political authoritarianism and social norms. Terrorism, quite frankly, cannot be understood by ignoring the context which produced it in the first place. Tausch's wide-angle approach which is both trans-disciplinary and combining both qualitative and quantitative methodologies speak to this more comprehensive approach to terrorism.

Our discussion on Tausch's approach covered three aspects. The first relates to what drives recruitment into terrorism organizations. Tausch explains radicalization as a result of both restrictive gender norms and Islamist shari'a ideology. Importantly, Tausch's approach provides us with connections between moderate and militant Islamists. Second, Tausch's scholarship provides deep insights into how anti-Semitism is deeply embedded in Islamist movements. Worryingly he notes that anti-Semitism is widespread. In the MENA region. For instance, 74 percent of the population display some bias against Jews. This could well be another path to recruitment into terrorist organizations. Third, what is to be done? In keeping with his wide-angled approach, Tausch calls for a deeper transformation of Muslim society whilst also acknowledging that Islamist ideology does not conform to the tenets of Islam as faith.

In the final instance, Professor Arno Tausch had made an indelible contribution to scholarship on terrorism and one which has deeply enriched the field.

References

Alexiev, A. (2005). Tablighi Jamaat: Jihad's stealthy legions. *Middle East Quarterly, 12*(1), 1–9.

Bayat, A. (2005). What is post-Islamism? *Islam Review, 9*(4), 10–25.

Block, L. (2005). Evaluating the effectiveness of French counterterrorism. *Terrorism Monitor, 3*(15), 13–23.

Freilich, J. D., Kerodal, A. K., & Galietta, M. (2015). Introducing the special issue on new directions in terrorism research. *International Journal of Comparative and Applied Criminal Justice, 39*(4), 57–81.

Hassan, K. (2005). Tablighis held in terror probe in Argentina. *The Daily Nation*. Retrieved August 4, 2022, from http://www.dailytimes.com.pk/default.asp?page=story_5-9-2005_pg1_4

Hindustan Times. (2021, December 12). BJP leader welcomes Saudi Arabia's 'gates of terrorism' order on Tablighi Jamaat. *Hindustan Times*. Retrieved August 4, 2022, from hindustantimes.com/india-news/bjp-leader-welcomes-saudi-arabia-s-gates-of-terrorism-order-on-tablighi-jamaat-101639318678335.html

Institute for Economics and Peace. (2020). *Global Terrorism Index 2020*. Institute for Economics and Peace.

Komecoglu, U. (2020). *Islamism, post-Islamism and civil Islam*. Retrieved February 26, 2020, from https://hudson.org/research/10032-islamism-post-islamism-and-civil-islam.

Laqueur, W. (2004). *Voice of terror: Manifestos, writings and manuals of Al Qaeda, Hamas and other terrorists from around the world and throughout the ages*. Reed Press.

Post, J. M. (2007). *The mind of the terrorist: The psychology of terrorism from the IRA to Al-Qaeda*. Palgrave Macmillan.

Rehman, S., & Askari, H. (2010). How islamic are islamic countries? *Global Economy Journal, 10*(2), 75–91.

Sageman, M. (2014). The stagnation. *Terrorism Research, Terrorism and Political Violence., 26*(4), 98–113.

Segell, G. (2021). The Abraham Accords: From extremist politics to diplomacy. In H. Solomon (Ed.), *Directions in international terrorism: Theories, trends and trajectories* (pp. 124–141). Palgrave Macmillan.

Solomon, H. (2013). *Jihad: A South African perspective*. Sun Media.

Solomon, H. (2017). *Hezbollah: The criminal syndicate, Middle East Institute occasional paper series*. Middle East Institute, New Delhi. Retrieved August 4, 2022, from mei.org.in/hezbollah-the-criminal-syndicate.

Solomon, H., & Tausch, A. (2020). *Islamism, crisis and democratization: Implications of the world values survey for the Muslim World*. Springer Nature.

Solomon, H., & Tausch, A. (2021). *Arab MENA countries: Vulnerabilities and constraints against democracy on the eve of the global COVID-19 crisis*. Springer Nature.

Tausch, A. (2001). *Globalization and European integration*. Nova Science Publishers.

Tausch, A. (2009). *What 1.3 billion Muslims really think: An answer to a recent Gallup study, based on the World Values Survey*. Nova Science Publishers.

Tausch, A. (2010). Globalisation and development: The relevance of classical "dependency" theory for the world today. *International Social Science Journal., 61*(202), 89–102.

Tausch, A. (2014). The new global antisemitism: Implications from the recent ADL-100 data. *Middle East Review of International Affairs, 18*(3), 34–54.

Tausch, A. (2015). Estimates on the global threat of Islamic state terrorism in the face of the 2015 Paris and Copenhagen attacks. *Middle East Review of International Affairs, 19*(1), 23–56.

Tausch, A. (2020). The return of religious anti-semitism? The evidence from world values survey data. In S. Hussein & A. Tausch (Eds.), *Islamism, crisis and democratization: Implications of the world values survey for the Muslim World* (pp. 145–167). Springer.

Tausch, A. (2021). Social attitudes fuelling islamist terrorism. In S. Hussein (Ed.), *Directions in international terrorism: Theories, trends and trajectories* (pp. 67–89). Palgrave Macmillan.

Tausch, A., Heshmati, A., & Karoui, H. (2015). The political algebra of global value change. *Society and Economy, 37*(4), 75–92.

Tausch, A., & Prager, F. (1993). *Towards a socio-liberal theory of world development*. Springer.

Tessler, M. (2015). *Islam and politics in the middle east: Explaining the views of ordinary citizens*. Indiana University Press.

Tibi, B. (2002). *The challenge of fundamentalism: Political Islam and the new world disorder*. University of California Press.

Tibi, B. (2012). *Islamism and Islam*. Yale University Press.

United Nations Development Program. (2022). *United Nations Social Norms Index 2022*. Retrieved August 4, 2022, from sdg.iisd.org/news/undp-gender-social-norms-index-finds-massive-bias-against-women.

Xulu, L. (2021). SA-based Islamic group rejects links to terrorism. *The Witness,* 20 December 2021. Retrieved August 4, 2022, from news24/com/witness/news/kzn/sa-based-islamic-group-rejects-links-to-terrorism-20211219.

Terror in the Twenty-First Century: The Modus Operandi of the Islamic State—What We Know and What We Can Learn from It!

Markus Reisner

Abstract The goals of the new terrorists are global and aim at a complete transformation. Developing effective counterstrategies will be one of the great challenges of the twenty-first century and of the current and future generations. The democratic constitutional state is called upon to prepare itself for the future challenges of supra-regional, global terrorism. This also includes dealing with the phenomenon of encountering an increasingly militarily active opponent. This in turn leads to the necessity of deploying military forces at home. National legislation must be subjected to critical examination and efforts must be made to adapt it to the new circumstances. In doing so, however, the democratic understanding of law and its basic pillars and goals, such as human dignity, freedom, justice, and legal security, must not be undermined. Can terrorist attacks be a case for military national defense? How should heavily armed military suicide bombers be dealt with? How to prevent one's own citizens or immigrants from slipping into extremism? How can people who have renounced terror be integrated? These are burning questions that need to be answered.

1 Introduction

Until the 1980s, many terrorist attacks were the work of small and regionally delimited revolutionary and national movements, often acting asymmetrically. Although these had become internationalized, the result of their attacks was usually manageable, and they could be kept in check by the police and military. This only changed when the motivation of terrorist groups shifted towards religious fanaticism. Bruce Hoffmann was one of the first to investigate this change on behalf of the American RAND Corporation (Hoffman, 2017, 83). In 1968, according to his assessment, of thirteen identifiable terrorist movements, not even one could be classified as religious. In 1995, 26 of 56 movements were religiously motivated. A

M. Reisner (✉)
Austrian Armed Forces, Vienna, Austria

G. Segell (ed.), *Development, Globalization, Global Values, and Security*,
https://doi.org/10.1007/978-3-031-24513-8_16

first high point that attracted global attention occurred in February 1993, when a terrorist attack was carried out by a previously unknown organization called Al-Qaida on the World Trade Center in New York (Townshend, 2005).

But things have changed since then. In his extensive studies, Arno Tausch has repeatedly shown us over the years that terrorist organizations are gaining increasing support, especially among the populations of the Arab World (Tausch, 2015). These developments are not diminishing; on the contrary, they are slowly but steadily increasing. There are ups and downs from year to year, but the presence of an increasing support for terrorist and Islamist ideas seems clear. As Arno wrote in one of his remarkable papers already in 2015: "According to the latest Arab Opinion Index, 3% of all Arabs now openly and strongly support ISIL (Daesh), and 2% support it to some extent. Support for ISIL (Daesh) declined since 2014, when 4% of all inhabitants of the Arab World had a very positive view of ISIL (Daesh) and a further 7% had a to some extent positive view of ISIL (Daesh)." (Tausch, 2015, 19).

The foundation of Islamist terrorist movements was often Salafist ideas. The ostensible goal was to return to the "true" teachings of the Koran, which these groups accepted as the only ones, and to wage war against all those who stood outside this form of Islam. In the process, the term jihad is repeatedly bandied about. While this word is often interpreted as the basic effort of every Muslim in his or her life, it is also seen as a guide to fight against everything un-Islamic. As an integral call to violence. The goal of Islamic terrorism is the establishment of a "true" Islamic state and social order. Violence is seen as the only way to bring the jihad to a successful end (Dietl et al., 2006, 135).

One also speaks of the "3–2-1 model" of jihad. One fights against three enemies (the "crusaders," i. e., Western societies; the "Jews," i. e. the state of Israel; the "stooges," i.e., Muslim governments or dictatorships) with two methods of attack (conquest and occupation of one's own Muslim countries and supra-regional export of terror) within the framework of a common global strategy of Islamism (as a so-called jihadist international).

The Islamists' fight is against "crusaders," "Jews" and "infidels" as well as their "henchmen" (that is also "apostate" Muslims). It is waged globally and includes physical and psychological attacks wherever they are deemed necessary. The strategy distinguishes between the fight against "near" and "far" targets. For example, the governments of the Arab states are to be overthrown first and only then the West is to be subjugated.

The leaders of Al-Qaeda and subsequently the Islamic State (IS) made sure that these messages were heard around the world. They were underpinned by spectacular terrorist attacks. These attacks and the resulting fear and uncertainty in many regions of the world led to the inspiration of further fighters. These were able to compete with the "infidels" in the Balkans from the mid-1990s, subsequently in the Caucasus, in Afghanistan from 2001 and in Iraq from 2003. The result was a breadth of operational experience that was exported to other regions and states, such as the Sahel region or North Africa. This is evident in different research, for example, Nasiri (2006, 34), Hegghammer (2010, 5), Elsässer (2005, 56) and Deliso (2007, 78).

Further germ cells of Islamist terrorism emerged there. The formation of these cells was not limited to the war zones, but also took place in the Western world. Since the violent death of Osama bin Laden, Al-Qaeda increasingly took a back seat as the inspiration, while IS and its (already killed) leader Abu Bakr al-Baghdadi took over the vanguard role (Tausch, 2021, 18). More and more Islamist terrorist groups pledged the Bai'a, the Islamic oath of allegiance, to the IS. In many Muslim countries, offshoots of the IS were formed, and the popularity of the group grew among the population (Tausch, 2021, 19). As in previous decades, Islamist terror did not stop at the gates of Europe. Here, it continues to pose enormous challenges for the police and military security forces (Morell, 2015, 34).

2 Terrorism as Part of Asymmetric or Irregular Warfare

Different countries, for example, USA, UK, Germany, but also international organizations for example the UN, NATO, EU offer different explanatory models for asymmetric or irregular warfare. A variety of terms exist on the subject, such as: Small War, Unconventional Warfare, Three Block War, fourth Generation War, Irregular Warfare, Subversive War, Insurgency, Guerilla War, Military Operations other than War, Low Intensity Conflict, Asymmetric War and finally Revolution, Partisan War and Terrorism. It should be noted that there is also no general agreement regarding the individual definitions of the terms mentioned. Here, too, the minds of science and research are divided. The author would like to point out this unsatisfactory circumstance by naming a range of terms.

The military, as an essential actor in the fight against terrorism, nowadays essentially distinguishes between five types of asymmetric or irregular warfare as examined by (Kiras & Jordan, 2008, 232). Important and in terrorism research, there is also the approach that terrorism should not be seen as a form of war but as a sui generis form of violence. The statement that terrorism is a form of military (in this case asymmetrical) warfare is mainly due to the approach of a historical study. Richard Clutterbuck, for example, does not make a categorical distinction between terrorism and guerrilla warfare, but uses the terms urban guerrilla warfare and urban terrorism largely synonymously (Clutterbuck, 1990).

Some of these are Coup d'état: The aim here is to bring about a turnaround by means of a targeted strike against a defined group for example government) that is as violent as possible. Terrorism: Small groups or cells with limited resources try to achieve a goal through the offensive spread of fear and terror. Revolution: A politically motivated group mobilizes large parts or a "critical mass" of an (oppressed) population with the aim of changing the situation. Insurgency: Based on a network of supporters and informants embedded in selective sections of the population, an attempt is made to achieve a specific goal. Civil War: Within a civil war, different groups fight for territory, recognition, and the legitimacy to govern a state in the future.

In addition, within each of these forms, the exercise of terror can represent a means (or strategy) of warfare. Considering the classification described above, the following can be said about the definition and concept: Terrorism is a form or strategy of asymmetric or irregular warfare in the modern military understanding. Herfried Münkler (2006) aptly summarizes this as follows:

"Terrorism is the practice of using violence by actors who are clearly inferior to their opponents in terms of resources and who want to achieve political goals through the psychological effects of physical violence. The term "ism" stands for the fact that these are not individual actions, but rather campaigns that are planned for a longer period, reflexively and strategically."

Among the most successful terrorist organizations of the early twenty-first century from a military point of view are Al-Qaeda (AQ) and the Islamic State (IS). While AQ operated mainly from behind the scenes, IS succeeded in bringing large areas under its de facto control for the first time. For a few years, it seemed as if the advance of IS could not be stopped for the time being. Currently, however, it seems as if the IS's days are numbered. This is mainly the result of a coordinated multinational effort by military intelligence services, coupled with the targeted deployment of special and security forces (Cockburn, 2015, 43).

In IS's heartlands in Syria and Iraq, defeat has followed defeat in recent years and its supporters seem to be fighting an increasingly hopeless battle. In fact, despite the successes of the intelligence services and security forces, IS still manages spectacular successes. Its offshoots in the Sahel (ISGS) and Afghanistan (ISIS-K) also remain active and carry out attacks. Ongoing analyses by relevant international military intelligence services indicate that an increased movement of fighters into the European area is still to be expected (Wood, 2018, 55).

The attacks that have taken place on European (and Austrian) soil in recent years may only have been a foretaste. It therefore seems appropriate to take an analytical look at the military combat techniques and tactics that these fighters might take with them to Europe. The following text thus primarily serves to understand a possible threat spectrum that can be brought to bear by a terrorist organization that is already militarily active in many areas of capability (in this text, the IS as an example). This in turn results in concrete deductions for military national defense, including the use of military intelligence services for the early detection of a possible threat. While the intelligence services' advance reconnaissance must be carried out in close cooperation with the police and the military, military national defense always remains the last resort in the case of a terrorist threat (Münkler, 2006).

3 The Roots of the Islamic State

When Abu Musab al-Zarqawi first thought about founding a new Islamist resistance and terrorist movement in Iraq at the end of 2005, he probably would not have suspected that his ideas would have such success. Previously an acolyte of Al-Qaeda in Iraq (AQI), he had broken away from it (despite the Islamic oath of allegiance to

Baiʿa) and laid the foundation for a new terrorist organization. Today we know this under the name Islamic State in Iraq and the Levant (ISIL) or simply Islamic State (IS) or, in Arabic, Dāʿish. In December 2006 (a few months after al-Zarqawi's violent death), the IS proclaimed an Islamic caliphate in Mosul, Iraq. Before the eyes of the American occupying forces and their Kurdish allies in northern Iraq. The latter could not explain why the US forces had not previously crushed the last pockets of resistance in Mosul in time. At the end of 2006, however, IS was still too weak to underpin its Islamic caliphate with actual gains in terrain. The necessary resources were still lacking. These first had to be procured and a functioning organization of fighters first had to be built up (Cockburn, 20).

Now, this article will not go into detail about the emergence of IS, but it is essential that today it is assumed that a not insignificant number of former Iraqi, militarily trained Sunni secret and intelligence officers of Saddam Hussein played a significant role in the emergence of IS. Many of the founding members of IS were also united by an ideological hatred of Iraq's Shiite majority, which has been politically dominant since the US invasion (Boucher, 2004).

It had been a declared goal of al-Zarqawi to instigate a civil war between Shia and Sunna. This was eventually waged with unusual ferocity just a few years after the international coalition invaded, often reducing the foreign occupiers to spectators of a bloody witch's cauldron. From its covert Sunni retreats in the west and north of Iraq, the IS expanded into Syria and opened the fight against the Alawite Assad regime there. Here, new opportunities arose to generate resources and, above all, weapons. Within a few months, they already controlled the Syrian provinces of Raqqa and Deir el Zour as well as their capitals.

From there, the IS finally returned strengthened to Iraq via the Syrian/Iraqi border near Al-Qaim. Welcome from the population of the cities of the so-called Sunni triangle such as Fallujah or Ramadi. The Sunnis had by now had enough of the oppression by Iraq's Shiite-dominated central government. After the IS had gained a foothold in the Iraqi provinces of Al-Anbar and Ninawa, coming from Syria, and had already taken control of major cities such as the afore-mentioned Fallujah in January 2014, the IS attacked the city of Mosul in June 2014 and took it in a coup d'état. Shiite-dominated security forces fled headlong, leaving behind equipment, devices, and ammunition. The rest is history. The world looked at Iraq and wondered: how could this have happened? (Clarke, 2019, 10).

4 The Secret of Extraordinary Success

The territorial conquest successes in Syria, the almost "peaceful" capture of the Sunni areas of western Iraq and the fall of the million-strong state of Mosul signify an enormous success for the IS. In December 2015, the IS was on the verge of capturing Aleppo, a city of millions, in Syria, and in Iraq its devastating attacks in Baghdad led to chaos and insecurity. While the international community disagreed on what to do, the US began sending troops back into Iraq. The task now was to prevent an imminent collapse of the Shiite Iraqi government.

Meanwhile, thousands of young men from all over the world flocked to Syria and Iraq to make sense of their lives: Fighting for Allah against the infidels or heretics who have fallen away from the faith (such as Alawites or Shiites). The IS used the next few months to exploit its victories for propaganda purposes, to secure conquered territories and to capture new ones. On January 7, 2015, IS sympathizers stormed the newspaper editorial office of Charlie Hebdo, and to this day Europe is experiencing a wave of terrorist attacks whose end is not yet in sight. The IS had achieved decisive successes in only a short time. It had managed to make many members of the Islamic Ummah identify with its goals and successes. The "accolade" for any terrorist organization. The ideas and ideals of the IS had become self-perpetuating (Clarke, 2019, 39).

Now one wonders how it could be that a few could become so many? How could it be that a dozen terrorists, some of them blood young, with Ak-47 assault rifles, a few pick-ups and a black banner bearing the Islamic creed and the seal of the Prophet (and thus representing familiar symbols of identification for the entire Islamic world), could achieve one success after another? Iraqi troops equipped with American M1 Abrams tanks, as well as Syrian regime forces with Russian-made tanks and armored personnel carriers, fled headlong at the sight of the black flags and abandoned their positions by the hundreds. Even the "Western" Turkish forces had to endure the loss of an entire company of Leopard 2 A4 tanks in the battle for the Syrian city of Al-Bab at the end of 2016. Not to speak of rival (Sunni but also Kurdish) rebel groups, which were virtually eliminated by IS fighters as they marched past. All these successes of the IS, however, reveal exact military patterns of action that, on closer inspection, could be taken from a military manual (Mumford, 2021, 21).

According to the NATO and EU Capability Hierarchy, modern armed forces are defined by mapping a specific number of capability areas (NATO, 2019). Modern armies are guided by these capabilities in the ongoing transformation of their armed forces. The Austrian Armed Forces are also constantly guided by these development and capability goals. These (listed by the author in the following without weighting according to priority) are referred to as: Intelligence and Reconnaissance (Inform), Leadership (Consult, Command & Control), Operational Preparation (Prepare), Impact and Engagement (Engage), Force Protection (Protect), Force Projection (Project) and Support and Sustainment (Sustain).

If a modern army possesses the characteristics of these capability areas, it is ultimately capable of the multi domain warfare, or of proactive cross-force command and control within the factors of force, space, time, and information. The author's thesis is that within a few years, the IS has managed to map all these military capability areas of modern armed forces. In varying degrees of quality and sophistication, but nevertheless complete and sustainable. If one looks at the combat techniques and operational tactics used by the IS, one concludes that a new quality of operational command by a terrorist organization has been and continues to be demonstrated here. This surprising quality of the IS's military action will now be examined in excerpts or by way of example (and considering the history of the IS's origins) (Clarke, 2019, 35).

5 Inform: Accurate Situational Awareness as the Basis for Targeted Action

Unlike any other regional terrorist organization in the Near and Middle East, IS succeeded from the very beginning in establishing and operating a widely ramified network of informers and spies. Against the background of the instigated intra-religious conflict between Sunnis and Shiites, the formerly dominant Sunni population strata developed a high willingness to cooperate with the newly emerged Islamist terrorist organization. This made it possible to carry out targeted attacks on the leaders of the Shiite-dominated security forces, especially at the beginning of the IS's emergence. These assassinations were also documented and disseminated to the media. The well-known and infamous multi-part video series called "Dancing Swords" is one example. The targeted killings not only intimidated the Iraqi security forces, but also drew the sympathy of the Sunni population. After all, it was the Shiite security forces who were feared for their attacks on Sunni civilians.

At the height of its success, that is the conquest of the city of Mosul with its millions of inhabitants, the IS knew through its system of informers exactly where to start (including a campaign of influence carried out by various media) and the city was in its hands within a few hours. Many of the IS's successes were therefore possible because its leadership knew exactly where to deploy its armed forces, which were still few at the beginning (Weiss 2015, 68).

6 Consult, Command, and Control: Tight Hierarchy Paired with Decentralized Order Fulfilment

It is currently assumed that the IS leadership cadre is not necessarily recruited from ideologically stable and strictly religious men, as was initially assumed. On the contrary, time and again, indications suggest that a conglomeration of different characters has come together. Be it former Sunni intelligence and secret service officers on the one hand or young technology freaks immigrating from the West on the other. It can rightly be assumed that at the core a manageable circle of thinkers carried out a forward-looking strategy development. In the beginning, the focus was on motivating the Sunni population to support them, and only then on the goal of steadily developing their own military capabilities.

Looking at the IS's fight in Syria in particular, it is striking that its attack targets were chosen wisely, especially at the beginning. The IS never tried to take over, but often filled a vacuum that had arisen or cleverly played rival rebel groups off against each other. This was accompanied by a centrally controlled media campaign of a quality that was previously unknown. The professional presentation of well-known IS propaganda formats such as Dabiq, Amak or Rumajah are examples of this. Some rebel groups (such as Jabhat an-Nusra, which existed until 2017) were much more successful than IS at the beginning, but their successes were not sufficiently exploited by the media. Soon, the first Syrian insurgents voluntarily joined the IS. Here, too, the IS skillfully used and mixed local and imported fighters. The self-proclamation of Abu Bakr al-Baghdadi as caliph and successor of the Prophet gave the IS overall strategy the necessary ideological and Islamist fine-tuning. Tight leadership, increasing military successes and ideological legitimacy gave the IS prestige and acceptance (Weiss 2015, 40).

7 Prepare: Forward Planning and Targeted Preparation

In the deployment of its forces, IS fighters differed fundamentally from other terrorist or rebel groups and regular armed forces operating in the space. Through the targeted selection of capable and experienced leaders, the individual military operations were planned and prepared in a considered manner. Based on a situation picture often generated over several weeks, precise plans were drawn up for necessary attacks. The operational command was divided into phases and precise roles were assigned to different available weapon systems. The reports of the Kurdish Peshmerga, who were repeatedly surprised by the attacks of the IS combat groups, are a particularly impressive example. The Zab River forms a natural border between the Iraqi provinces of Ninawa and Arbil.

Despite a security line of Kurdish forces that existed before 2003, the IS repeatedly succeeded in crossing the river, infiltrating fighters undetected and destroying entire Kurdish bases down to the last man. In addition, there were advances deep into the Kurdish hinterland where, among other places, attacks were carried out in the informal Kurdish capital Arbil. The situation was similar in Baghdad, where the IS repeatedly succeeded in carrying out devastating car bomb attacks despite the strong presence of the Iraqi security forces. Here, too, the IS had relied on the population of the Sunni neighborhoods. Through careful planning, targeted analysis of weak points and preparation based on sympathizers, the IS succeeded in achieving successive spectacular successes (Weiss 2015, 131 and Supplemented by the author's own observations during stays in Mosul, Kirkuk, Makhmur and Arbil in 2008).

8 Engage: Initiative Maneuver Warfare and High Flexibility

It seems that the military planners of the IS have carefully studied the campaigns of the Umayyad dynasty in the eighth century. They also took inspiration from the campaigns of legendary historical equestrian peoples such as the Avars, Huns, or Mongols. Their campaigns were characterized by the fact that, supported on horseback, they conquered large areas in a short time. Heavily armored armies of knights from the West were skillfully outmaneuvered or covered with arrows from a distance until they were worn down. The Umayyads had advanced along the North African coast, crossed the Mediterranean and invaded Spain. There they ruled until the eleventh century.

Like their Arab ancestors, the IS was characterized from the beginning by highly mobile operations. Relying on light vehicles (for example, primarily pick-ups) with medium armament, the open deserts of Iraq and Syria in particular offered the possibility of rapid advance. After the success of the covert operations and with the availability of captured weapons and vehicles, there was a shift to offensive operations. The first step was to capture every enemy position, which was easy to conquer in the vastness of the desert areas. The enemy bases were first encircled, then fired upon with large caliber weapons and finally stormed.

On the one hand, tactically insignificant targets were attacked and on the other hand, strategic nodes were targeted. The capture of each base spreads like wildfire among the Syrian Alawite, Iraqi Shiite or Kurdish defenders. This led them to abandon their bases as soon as the dust clouds of an approaching unknown convoy of vehicles could be seen on the horizon. The flexible, highly mobile deployment of its forces allowed IS to quickly exploit identified vulnerabilities and shift forces at will. It was only with the launch of the US Operation Inherent Resolve from autumn 2014 and the Russian intervention in Syria from September 2015 that this approach of the IS was made considerably more difficult by the start of targeted air strikes (Mumford, 81).

9 Protect: Protection Against Modern Weapon Systems

It would be easy to attribute IS's success exclusively to its highly mobile operational command. This would also imply that its opponents always took flight during IS attacks. The opposite was often the case. The Iraqi and Kurdish garrisons desperately fought to the last in their isolated and enclosed bases. They knew what would await them if they were captured by the IS. To this day, IS's penetrating "precision weapons" are vehicles loaded with explosives, fitted with makeshift armor and driven by suicide bombers. So-called Suicide Vehicle Born Improvised Explosive Devices (SVBIED) were used from the beginning, and in all conceivable versions. Initially, they were used to blow up bases completely or to make a breach. If the breach was insufficient, the next vehicle was brought in and detonated. Short-range anti-tank missiles or heavy machine guns of the defenders were countered with appropriate armor. Or the IS simply brought in captured armored personnel carriers, armored vehicles or tractors. The vehicles were each loaded with up to several hundred kilograms of different explosives and blasting agents. Each detonation therefore brought with it a devastating radius of destruction.

Far beyond the effective range of the defenders' shoulder-launched anti-tank weapons. The only effective response turns out to be long-range anti-tank guided weapons, deep armored trenches or barriers, or . . . as a last resort . . . flight from the attacker. To date, IS has continued to refine the use of its SVBIED. Hundreds (!) of SVBIEDs were used in the battle for Mosul. In the process, weak points in the Iraqi forces' security lines were reconnoitered by means of mini-drones and SVBIEDs were targeted at them. Armored trenches were countered at favorable points with makeshift armored construction machines. If a trench could be filled in, swarms of SVBIED followed. Finally, the drones used for reconnaissance were loaded with explosives and used. Either by deliberate crash or by dropping explosives. There seems to be no shortage of drivers and suicide bombers to this day.

It seems that, in contrast to the Western soldier, they are much less attached to their existence in this world. For them, death for Allah represents the fullfilment of their greatest goal. In the battle for Mosul, IS even took to hiding SVBIED in garages and activating them only after the Iraqi forces had passed. Thus, garage doors repeatedly opened in the back of the Iraqi forces and unloaded their deadly cargo. Or civilians were forced to ride along for "protection" against US air strikes. Such operational tactics led to Iraqi special and security forces (for example the Counter Terrorism Service, CTS or Jihaz Mukafahat al-Irhab) suffering up to 50% casualties in the battle for Mosul. With simple means, coupled with ideological brainwashing (and probably also coercion), the IS had succeeded in developing an extremely successful simple weapons system (Weiss, 114).

10 Project: Efficient and Highly Mobile Use of Low Forces

Despite successful local recruitment, the immigration of thousands of fighters from abroad and (especially at the beginning) a high level of sympathy among the Sunni population, the IS was faced from the beginning with the challenge of holding and securing the captured territories. The Iraqi Mesopotamia, but also Syria, is characterized by a concentration of its inhabitants in urban areas, but in order to hold these, it needed forces. At no time, however, were these available to the IS in sufficient numbers. In many places, the solution was both simple and terrible. By using the most brutal methods, such terror was exercised that often the mere mention of IS terrified people.

This too is not new in warfare. Recall again the Hun, Mongol, and finally Ottoman advances into Europe. The Turkish Janissaries were preceded by the Sipahi, the Turkish cavalry. And the Akinci spread fear and terror in front of them.

They often acted so brutally that the local population fled at the mere rumor of their approach. Hardly any prisoners were taken. They would have needed to be guarded and fed and, what is more, they were infidels. This categorization in turn reduced the inhibition to kill. It was mainly young boys (to train them as janissaries) and women (to satisfy sexual needs) who were abducted. The Huns and Mongols had acted similarly before.

The IS copied this tactic almost perfectly. Iraqi, Syrian, or Kurdish soldiers were executed by the thousands, some of them with select and carefully documented cruelty, young boys trained to be fighting machines and women abducted as needed. Any rebellion by the (now increasingly Sunni) population was nipped in the bud with uncompromising cruelty. People were hanged, burned, skinned, stoned, run over, beheaded, shot, hacked to death, beaten to death and crucified. Behind this bestiality of the IS, however, was above all the sober calculation to nip all resistance in the bud by means of terror. Islamist ideology provided the necessary legitimacy. In this way, the IS succeeded in dominating large areas with only a few forces (Clarke, 39).

11 Sustain: Holding the Won Terrain to the Last

In addition to the difficulty of holding the terrain gained, supplying one's own forces also posed a great challenge. Moreover, it was necessary to give the population in the conquered areas the feeling of being able to recognize significant progress in a caliphate after years of war. The use of terror alone is not enough to keep a population calm. The IS used all possible means to achieve this. Taxes were collected from the population of the conquered areas and efforts were made to repair destroyed infrastructure as quickly as possible. Local sympathizers were entrusted with leadership tasks and used to recruit more fighters.

All measures were accompanied by a comprehensive media campaign. This attracted more fighters to the supposed paradise from afar. A flourishing trade in oil and various refinery products developed. Columns of hundreds of fully loaded tankers traveled across all borders for a long time. Tolerated by prominent neighboring states. Even art treasures were sold on a grand scale. Weapons and ammunition of all calibers were traded between the opposing groups. Time and again, weapons systems supplied to moderate rebel groups from abroad ended up with the IS. But the IS also went one step further. It managed to set up an extensive weapons and ammunition production. Including precise manufacturing and production specifications to maintain the necessary quality standards. Everything from (reloaded) infantry ammunition to medium-range rockets or explosive devices for drones was manufactured.

In the conquered territories, on the other hand, defense measures were taken immediately. Thus, even in small villages, extensive underground bunkers were built, water and basic foodstuffs were stockpiled, and ammunition supplies were stored. During the recapture by the Iraqi forces, these installations, which were often booby-trapped and difficult to reconnoiter from the air, posed a great challenge. Often, fierce battles were fought for hours over individual blocks of houses. In addition, civilians were taken hostage. This made the targeted use of artillery and air strikes difficult or impossible. Every civilian killed by the international coalition forces showed the limits of precise weapons deployment. From the beginning, IS had prepared for a fight in the areas it had conquered. Thus, to this day, it manages to hold out in towns and villages that have already been destroyed and to prepare for a final battle (Bob & Teller, 2018).

12 What Does This Mean for the World?

In summary, IS can be seen as an example of how it is possible in a short period of time to sustainably conquer a space and hold it militarily (at least temporarily). The final defeat of IS in the coming years will depend on the sustained combined deployment of military intelligence and secret services together with security forces. But what already remains is its legacy and the fact that thousands of sympathizers have learned to be effective in all areas of capability of modern armed forces. Although the IS never had an air force or even a navy, and even its land forces were never structurally comparable to regular armed forces, it nevertheless managed to take possession of a large space.

To this day, the terror perpetrated in its name radiates all the way to Europe. Even in the Sahel, Libya, Afghanistan, and the Philippines, attacks are being carried out in its name, or new nuclei are trying to form. An analysis of its combat techniques and operational tactics nevertheless shows familiar patterns of action, and this means that organizations like the IS will in any case remain combatable and neutralizable in the future. However, if one looks at the basic principles of international humanitarian law (for example, proportionality in the use of means, the principle of distinction or humanity in war), one realizes that the price the international community has to pay for such successes is becoming increasingly higher. With the emergence of IS groups in the Sahel, Afghanistan, the Philippines and Libya, the application of its combat techniques and operational tactics can also be seen there. And even the attacks in Europe clearly show the acquired military skills of covertly operating IS cells.

The neutralization of the attackers on November 18, 2015, in St. Dennis after the Paris attacks is a cautionary example. Only after hours of fighting, five injured police officers, one killed police dog, the use of 3200 rounds of ammunition and 40 offensive and two defensive grenades as well as twenty rifle grenades by French police special forces of RAID (Recherche, Assistance, Intervention, Dissuasion) was it finally possible to eliminate the attackers. A first bitter foretaste of what may be to come (Clarke, 105).

13 Necessary Derivations for the Use of Military (Intelligence) Services

The example of IS shows that the quality of planning and the means used (explosives, weapons and fighters) as well as the terrorist capability spectrum are constantly increasing. An IS fighter who has fled (or returned) to Europe from Iraq and Syria is very likely to have military knowledge which, in its scope and sophistication, can pose an enormous challenge to police intelligence services and security forces. When drawing up threat analyses, preference must therefore always be given to the whole-of-government approach. Police and military intelligence services must continuously coordinate their reconnaissance results. Newly identified tactics and combat techniques must be incorporated into the education and training of police and military personnel. If the terrorist increasingly uses military capabilities, the military must be involved in the analysis and in the development of possible defense strategies (Münkler, 2006).

The search for the roots of terrorism starts with terminology. In order to be able to describe the phenomenon of terrorism correctly, the relevant terms must first be defined from today's perspective. The author understands terrorism as a form of so-called asymmetric or irregular warfare. It should be noted here that the opponents (conventional as well as unconventional) of any military conflict strive to achieve asymmetry. This type of warfare is not new in essence but has changed in quality and form over the centuries. To date, there is no internationally recognized definition for asymmetric or irregular warfare or for the term terrorism itself. As early as 1977, the internationally recognized terrorism expert Walter Laqueur stated that "... a general definition of terrorism does not exist and will not be found in the near future" (Daase & Spencer, 2009).

14 Summary and Outlook

As shown in this essay, there are different ways of approaching the phenomenon of terrorism. Many scholars try to explain terrorism from a social science perspective. The author has taken a historical approach. Furthermore, the problems that arise from a legal point of view (for example, on the subject and classification within international humanitarian law) should be pointed out.

Not least the interventions of international forces in Afghanistan, Iraq and Libya and there ongoing presence in the Sahel led to the fact, that more and more states are becoming the focus of Islamist terror. In addition, many thousands of young men, but also women, have gone to Syria and Iraq to fight side by side with their brothers in faith. Quite a few of them returned and now pose an internal European threat. The Islamists' fight is now directed against their own fellow believers. Thus, Shiites are considered heretics by the Sunni followers of IS and their death is seen as the only true goal. This new hostility has led to hundreds of thousands of deaths in the Middle

East so far. Many times, more than in the terrorist attacks carried out so far against Western institutions. Islamism, with its radical objectives, must be seen as a strategic challenge.

The ideas of the Islamists fundamentally contradict the ideas of modern society. Finding a compromise here through negotiation and compromise does not seem possible, at least at the present time. Terrorist groups nowadays operate increasingly decentral and often have an inspirational effect, that is they thus find imitators and followers. Once again, we should refer to the strategy of IS, which wants to win believers all over the world for its idea of Islamism. A combination of controlled and guided terrorism and so-called franchise terrorism is developing. Previously completely blameless citizens (with or without a migration background) carry out devastating attacks on their own initiative and refer to ideological-ideological, religious terrorist movements (Todenhöfer, 2015). The future of counterterrorism is full of challenges. We must be ready to recognize them, assess them and draw the right conclusions. Before it is too late.

References

Bob, Y. J., & Teller, N. (2018). *Ex-CIA official warns of Iran, ISIS drones armed with chemical weapons*, January 20, 2022. Retrieved August 4, 2022, from https://www.jpost.com/Middle-East/Ex-CIA-official-warns-of-Iran-ISIS-drones-armed-with-chemical-weapons-617132.

Boucher, R. (2004). *Spokesman: Foreign Terrorist Organization: Designation of Jama'at al-Tawhid wa'al-Jihad and Aliases*. United States Department of State, January 20, 2022. Retrieved August 4 2022 from https://2001-2009.state.gov/r/pa/prs/ps/2004/37130.htm.

Clarke, C. P. (2019). *After the Caliphate: The Islamic State & the Future Terrorist Diaspora.* Cambridge.

Clutterbuck, R. (1990). *Terrorism and Guerrilla Warfare*. Bloomsbury.

Cockburn, P. (2015). *The Rise of Islamic State: ISIS and the New Sunni Revolution*. Palgrave.

Daase, C., & Spencer, A. (2009). Terrorismus. In C. Carlo Masala, F. Sauer, & A. Wilhelm (Eds.), *Handbuch der Internationalen Politik*. Wiesbaden.

Deliso, C. (2007). *The Coming Balkan Caliphate – The Threat of Radical Islam to Europe and the West*. Westport.

Dietl, W., Hirschmann, K., & Tophoven, R. (2006). *Das Terrorismus-Lexikon*, Täter, Opfer. Hintergründe.

Elsässer, J. (2005). *Wie der Dschihad nach Europa kam – Gotteskrieger und Geheimdienste auf dem Balkan*. Wien.

Hegghammer, T. (2010). The Rise of Muslim Foreign Fighters – Islam and the Globalization of Jihad. *International Security, 35*(3), 5.

Hoffman, B. (2017). Inside terrorism, Revised and expanded edition. .

Kiras, J. D., & Jordan, D. (2008). Irregular Warfare. In J. D. Kiras & D. Jordan (Eds.), *Understanding modern warfare*. Macmillan.

Morell, M. (2015). *The Great War of our Time. The CIA's fight against Terrorism from Al Qa'ida to ISIS*. Harper Collins.

Mumford, A. (2021). *The West's War Against Islamic State: Operation Inherent Resolve in Syria and Iraq*. Palgrave.

Münkler, H. (2006). *Die Strategie des Terrorismus und die Abwehrmöglichkeiten des demokratischen Rechtsstaats*. Retrieved August 4, 2022, from https://edoc.bbaw.de/frontdoor/index/index/docId/55.

Nasiri, O. (2006). *Mein Leben bei al-Quaida – Die Geschichte eines Spions*. München Verlag.
NATO. (2019). *Consultation, Command and Control Board (C3b) C3 Taxonomy Baseline 3.1, Capability Hierarchy*. NATO STO Office.
Tausch, A. (2015). The fertile grounds for ISIL terrorism, *Telospress* (2015), January 20, 2022. Retrieved August 4, 2022, from http://www.telospress.com/tag/telos-171/
Tausch, A. (2021). *The future of the Golf Region – Value Change and Global Cycles*. Golf Studies 2. Springer.
Todenhöfer, J. (2015). *Inside IS – 10 Tage im „Islamischen Staat" (München), 5. –Pankratz, T. (2015) Islamistischer Terror – eine globale Bedrohung auch für Österreich?, Tagungsband 2. ALES-Tagung, January 20, 2022*. Retrieved August 4, 2022, from https://www.manz.at/list.html?isbn=978-3-86676-330-2
Townshend, C. (2005). *Terrorismus – Eine kurze Einführung (Ditzingen)*. Retrieved August 4, 2022, from https://journals.sagepub.com/doi/10.1177/0047117808100615, 99.
Wood, G. (2018). *The way of the strangers: Encounters with the Islamic State*. Sage.

Printed in the USA
CPSIA information can be obtained
at www.ICGtesting.com
CBHW052335111124
17272CB00003B/159

9 783031 245121